PROGRESS AND ITS PROBLEMS

PROGRESS and Its PROBLEMS

Towards a Theory of Scientific Growth

LARRY LAUDAN

UNIVERSITY OF CALIFORNIA PRESS
Berkeley Los Angeles London

University of California Press
Berkeley and Los Angeles, California
University of California Press, Ltd.
London, England

First Paperback Printing 1978
ISBN: 0-520-03721-9
Library of Congress Catalog Card Number: 76-24586
Printed in the United States of America

4 5 6 7 8 9

To Rachel, Heather & Kevin—
fellow pilgrims

Contents

Preface ix

Prologue 1

Part One: *A Model of Scientific Progress*

1. THE ROLE OF EMPIRICAL PROBLEMS 11

The Nature of Scientific Problems—Empirical Problems—Types of Empirical Problems—The Status of Unsolved Problems—The Nature of Solved Problems—The Special Role of Anomalous Problems—Converting Anomalies to Solved Problems—The Weighting of Empirical Problems—Theory Complexes and Scientific Problems

2. CONCEPTUAL PROBLEMS 45

The Nature of Conceptual Problems—The Sources of Conceptual Problems—The Relative Weighting of Conceptual Problems—Summary and Overview

3. FROM THEORIES TO RESEARCH TRADITIONS 70

Kuhn's Theory of Scientific "Paradigms"—Lakatos' Theory of "Research Programmes"—The Nature of Research Traditions—Theories and Research Traditions—The Separability of Theories from Research Traditions—The Evolution of Research Traditions—Research Traditions and Changes in Worldviews—The Integration of Research Traditions—"Nonstandard" Research Traditions—The Evaluation of Research Traditions—Adhocness and the Evolution of Research Traditions—Anomalies Revisited—Summary: A General Characterization of Scientific Change

4. PROGRESS AND REVOLUTION 121

Progress and Scientific Rationality—Scientific Revolutions—Revolution, Continuity, and Commensurability—Non-Cumulative Progress—In Defense of "Immature" Science

Part Two: *Applications*

5. HISTORY AND PHILOSOPHY OF SCIENCE 155

The Role of History in the Philosophy of Science—The Role of Norms in the History of Science—Rational Appraisal and "Rational Reconstruction"

6. THE HISTORY OF IDEAS 171

Disciplinary Autonomy and the History of Ideas—Ideas and their Problem Contexts—The Aims and Tools of Intellectual History—Problem Solving and Nonscientific Research Traditions—The Indispensability of History for Theory Appraisal

7. RATIONALITY AND THE SOCIOLOGY OF KNOWLEDGE 196

The Domain of Cognitive Sociology—The Theoretical Foundations of Cognitive Sociology—Conclusion

Epilogue: Beyond *Veritas* and *Praxis* 223

Notes 227

Bibliography 247

Index of Names 255

Preface

It has been my good fortune to have been student or colleague to many of the scholars whose work has done much to shape the character of contemporary history and philosophy of science: C. G. Hempel, T. S. Kuhn, Gerd Buchdahl, Paul Feyerabend, Karl Popper, Imre Lakatos and Adolf Grünbaum have all left their mark on the eclectic doctrines that make up this essay. If the book is persistently critical of some of their ideas, it is because healthy disagreement (unlike imitation) is the deepest sign of an abiding admiration. Unfortunately, it is no longer possible for me to acknowledge specifically how much my approach to science owes to each of these thinkers; but the collective debt is enormous. What originality there may be in this essay derives almost entirely from the insights (and in some cases the pregnant confusions) to be found in their writings.

Other forms of indebtedness are, however, easier to localize. Research grants from the National Science Foundation, from the German Fulbright-Kommission and from the University of Pittsburgh provided the release time necessary to undertake this project. The hospitality of the University of Konstanz provided a congenial atmosphere for putting to paper ideas which had been percolating in my seminars since 1970. Cindy Brennan and Karla Goldman provided yeoman service in preparing the manuscript. Specific portions of earlier drafts of this essay have been profitably discussed with A. Grünbaum, D. Hull, J. E. McGuire, K. Schaffner, M. J. S. Hodge, M. and R. Nye, I.

Mitroff, P. Machamer, N. Rescher, R. Creath, A. G. Molland, S. Wykstra, F. Kambartel, J. Mittelstrass, P. Janich, and J. M. Nicholas. The book would be far more flawed than it is without their criticism and suggestions. My greatest debt, however, is to Rachel, whose patience, critical sense, and unflagging encouragement sustained this project through its difficult incubation period.

June, 1976

Prologue

We must explain why science—
our surest example of sound knowledge—
progresses as it does, and we first must find out how,
in fact, it does progress. T. S. KUHN (1970), p. 20

Epistemology is an old subject; until about 1920, it was also a great one. What produced the change was a confluence of three quite separate developments, each of which effected a profound transformation in the study of knowledge. There was, first of all, the crisis produced by the realization that knowledge was neither as certain nor as incorrigible as thinkers since Plato and Aristotle had presumed it to be. There was, secondly, the increasing professional insularity of academic philosophers, and their related conviction that disciplines such as psychology and sociology, which had played a major role in earlier epistemological theories, had no interesting insights to offer. (This insularity was further promoted by the guileless duplicity of scholars in other fields, who were all too prepared to bequeath "the problem of knowledge" to the professional philosophers.) There was, finally and catastrophically, a growing tendency (especially in the English-speaking world) to imagine that one could grapple with the nature of knowledge while remaining blissfully ignorant of its best extant example—the natural sciences.

Despite the attempted appropriation of epistemological issues by the professional philosophers, many of the classical questions about the nature of scientific knowledge still remain of broad, general interest: Does science progress? Are our ideas about nature really worthy of credence? Are some beliefs about the world more rational than others? Issues such as these go well beyond the bounds of specialized disciplinary monopolies. They do so in large part because most people in the West draw the bulk of their beliefs about nature, and even about themselves, from the corpus of science. Without Newton, Darwin, Freud, and Marx (to mention only the more obvious), our picture of the world would be vastly different from what it is. If science is a rationally well-founded system of inquiry, then it is only right and proper that we should emulate its methods, accept its conclusions, and adopt its presuppositions. If, however, science is predominately irrational, then there is no reason to take its knowledge claims any more (or less) seriously than we take those of the seer, the religious prophet, the guru, or the local fortuneteller.

For a long time, many have taken the rationality and progressiveness of science as an obvious fact or a foregone conclusion, and some readers will probably still think it bizarre to believe that there is any important problem to be solved here. Although this confident attitude has been almost inescapable given the cultural biases in favor of science in modern culture, there have been a number of recent developments which bring it into serious question:

1. Philosophers of science, whose primary aim is to define what rationality is, have generally found that their models of rationality find few, if any, exemplifications in the actual process of scientific activity.[1] If we accept the claim made on behalf of these models to the effect that they define rationality itself, then we seem forced to view virtually all science as irrational.

2. Attempts to show that the methods of science guarantee it is true, probable, progressive, or highly confirmed knowledge—attempts which have an almost continuous ancestry from Aristotle to our own time—have generally failed,[2] raising a distinct presumption that scientific theories are neither true, nor probable, nor progressive, nor highly confirmed.

3. Sociologists of science have been able to point to several episodes in the recent (and distant) past of science which *seem* to reveal many nonrational, or irrational, factors decisively involved in scientific decision making.[3]

4. Some historians and philosophers of science (e.g., Kuhn and Feyerabend) have argued, not merely that certain decisions between theories in science *have been irrational,* but that choices between competing scientific theories, in the nature of the case, *must be irrational.*[4] They (especially Kuhn) have also suggested that every gain in our knowledge is accompanied by attendant losses, so that it is impossible to ascertain when, or even whether, we are progressing.[5]

The skepticism to which such conclusions point has been reinforced by the general arguments of cultural relativism to the effect that science is just one set of beliefs among many possible ones, and that we in the West venerate science, not because it is more rational than its alternatives, but simply because we are a product of a culture that has traditionally set great store by science. All systems of belief, including science, are seen as dogmas and ideologies, between which objective, rational preference is impossible.

Confronted by the acknowledged failure of the traditional analysis to shed much light on the rationality of knowledge, three alternatives seem to be open to us:

1. We might continue to hope that some as yet undiscovered minor variation in the traditional analysis will eventually clarify and justify our intuitions about the cognitive well-foundedness of science and thus prove to be a worthy model of rationality.

2. We might, alternatively, abandon the search for an adequate model of rationality as a lost cause, thereby accepting the thesis that science is, so far as we know, blatantly irrational.

3. Finally, we might begin afresh to analyze the rationality of science, deliberately trying to avoid some of the key presuppositions which have produced the breakdown of the traditional analysis.

Enormous efforts have been devoted, particularly in the last decade, to the pursuit of strategies (1) and (2). Philosophers of science, by and large, have taken the first option. Thus, Lakatos asks, "What are the *minimum* changes needed in the Popperian analysis of science to enable it to solve the problem of rationality?"[6] Salmon asks, "What are the *minimum* adjustments needed in Reichenbach's theory to square it with scientific practice?" Hintikka poses the question, "What kind of *tinkering* with Carnap's inductive logic will make it relevant to scientific testing?" While one admires the tenacity and ingenuity illustrated by proponents of this approach, the results are not, on the whole, very encouraging. Most of the difficulties which stood in the way of a Popper, a Carnap, or a Reichenbach still remain obstacles for their latter-day disciples.[7]

The second option has proved more popular with historically oriented thinkers. Thus, both Kuhn and Feyerabend conclude that scientific decision making is basically a political and propagandistic affair, in which prestige, power, age, and polemic decisively determine the outcome of the struggle between competing theories and theorists. Their mistake seems to be one of jumping to a premature conclusion. They start from the premise that rationality is exhaustively defined by a certain model of rationality (each of them takes Popper's model of falsifiability as the archetype). Having observed, quite correctly, that the Popperian model of rationality will do scant justice to actual science, they precipitately conclude that science must have large irrational elements, without stopping to consider whether some richer and more subtle model of rationality might do the job.

Because the one option seems unpromising and the other premature, I am inclined to think that we should consider pursuing the third strategy. Let us drop some of the traditional language and concepts (degree of confirmation, explanatory content, corroboration and the like), and see if a potentially more adequate model of scientific rationality begins to emerge. Let us see whether, by asking anew some of the elementary questions about science, we cannot get a slightly different perspective on scientific knowledge.

In what follows, I shall attempt to trace out the consequences of the view that science fundamentally aims at the solution of problems. Although the view itself is commonplace, very little attention has been given to exploring it in detail. What the different types of problems are, what makes one problem more important than another, the criteria for counting something as an adequate solution, the relation of nonscientific problems to scientific ones; none of these issues has been addressed in the detail it demands. To anticipate some of my conclusions, I propose that the rationality and progressiveness of a theory are most closely linked—not with its confirmation or its falsification—but rather with its *problem solving effectiveness.* I shall be arguing that there are important *nonempirical,* even "*non-scientific*" (in the usual sense), factors which have—and which should have—played a role in the *rational* development of science. I shall suggest, further, that most philosophers of science have mistakenly identified the nature of scientific appraisal, and thereby the primary unit of rational analysis, by focussing on the individual theory, rather than on what I call the *research tradition.* This study will show, moreover, that we need to distinguish between the *rationality of acceptance* and *the rationality of pursuit* if we are to make any progress at reconstructing the cognitive dimensions of scientific activity.

My basic strategy in what follows will involve the blurring, and perhaps the obliteration, of the classical distinction between scientific *progress* and scientific *rationality.* These two notions, both central to any discussion of science, have often seemed at cross purposes. Progress is an unavoidably *temporal* concept; to speak about scientific progress necessarily involves the idea of a process occurring through time. Rationality, on the other hand, has tended to be viewed as an atemporal concept; it has been claimed that we can determine whether a statement or theory is rationally credible independently of any knowledge of its historical career. Insofar as rationality and progressiveness have been linked at all, the former has taken priority over the latter—to such a degree that most writers see progress as *nothing more than* the temporal projection of a series of individual rational choices. To be progressive, on the usual view, is to adhere to a series of increasingly rational beliefs. I am deeply troubled by the unanimity with which philosophers have made progress *parasitic* upon rationality. In part, my worry arises from a concern that it involves explaining something which can be readily understood (progress) in terms of something else (rationality) which may be far more obscure. More serious, however, is the absence of any convincing argument as to why we should explicate our concept of progress in terms of rationality. The two concepts are doubtless related, but not necessarily in the manner usually supposed.

It will be the assumption here that we may be able to learn something by inverting the presumed dependency of progress on rationality. I shall try to show that we have a clearer model for scientific progress than we do for scientific rationality; that, moreover, we can define rational acceptance in terms of scientific progress. In a phrase, my proposal will be that *rationality consists in making the most progressive theory choices,* not that progress consists in accepting successively the most rational theories. This inversion of the usual hierarchy offers some insights into the nature of science which tend to elude us if we preserve the traditional relation between progress and rationality.

Another of the chief obstacles to the development of a theory of scientific progress has been the universal assumption that progress can occur only if it is *cumulative,* that is, if knowledge grows entirely by accretion. Because there are grave difficulties, both historically and conceptually, with the progress-by-accretion view, I propose a definition of scientific progress which does not demand cumulative development.

In order for the ambitions of this enterprise to be brought to fruition, and to prevent its being misconstrued, two key points must be stressed. First, the term "progress" has many *emotive* overtones deeply rooted in the subjective intuitions of both friends and critics of science. The object of this work is not to exploit that emotiveness, but rather to offer objective criteria for determining when progress has occurred. In too many discussions of progress, insufficient attention has been given to separating out the question of what progress is from the question of its moral and cognitive desirability. Any adequate theory of progress must make such a distinction as sharply as possible. There is a second crucial ambiguity in normal usages of "progress" which must also be noted. Specifically, it is commonplace to speak of progress, meaning an improvement in the material or the "spiritual" conditions of life. Although that sense of progress is unquestionably important, I shall say virtually nothing about it in this essay. My exclusive preoccupation will be with what I call *"cognitive progress,"* which is nothing more nor less than *progress with respect to the intellectual aspirations of science.* Cognitive progress neither entails, nor is it entailed by, material, social, or spiritual progress. These notions are surely not altogether disconnected, but they do refer to very different processes, and, at least for purposes of the present discussion, should be sharply distinguished.

One final point is in order. Previously, too many discussions of scientific rationality and progress have been both uninformed by, and inapplicable to, the actual course of the evolution of science. The various well-known philosophical models of rationality have been shown to be inapplicable to most of those cases in the history of science where, at least intuitively, we are convinced that sensible, rational choices were being made. Without assuming that whatever science does is, by definition, rational, we should nonetheless be able to demand of any model of science that it substantially "fit" the actual course of scientific change. Accordingly, historical cases and episodes will be used extensively in this essay; these are intended not merely to *illustrate* my philosophical claims, but also to *test* them. If the model under discussion here fails to illustrate the manner in which scientific decision making has actually worked (at least some of the time), then it will have failed entirely in its ambitions.

Because of the unusually heavy weight attached in this approach to historical material—material which some philosophers deem to be absolutely irrelevant to epistemology—I shall also briefly discuss the general question of the bearing of descriptive data (such as history) on a normative theory (such as a model of scientific rationality).

Part One of the following study articulates a model of scientific progress and rationality, and exhibits how that model, for all its evident incompleteness, avoids many of the paradoxes which previous models have generated, and makes some sense of the historical data. Part Two examines the ramifications of that model for a variety of intellectual inquiries, ranging from the history of ideas to the history and philosophy of science and the sociology of knowledge.

It has not been possible for me to explore all the issues concerned with scientific progress in the detail which they deserve. For that failure, I can only ask the reader's mercy. This is not, nor is it intended to be, a finished piece of work. At many points, argument sketches pass for arguments and plausible intuitions are invoked where, ideally, explicit doctrines are called for. A great deal remains to be said on all the matters I address. But the study of rational knowledge and its growth, like knowledge itself, is a cooperative venture of a community of minds. My purpose is merely to offer a fresh perspective on some problems which have preoccupied reflective people for a very long time.

Part One

A Model of Scientific Progress

The activity of understanding
is, essentially, the same as
that of all problem solving. K. POPPER (1972), p. 166

Chapter One
The Role of Empirical Problems

Problem formulation in science is to be understood by looking at the continuity of the whole stream of scientific endeavor. H. SIMON (1966), p. 37

Science is essentially a problem-solving activity. This anodine bromide, more a cliché than a philosophy of science, has been espoused by generations of science textbook writers and self-professed specialists on "*the* scientific method." But for all the lip service which has been paid to the view that science is fundamentally the solving of problems, scant attention has been paid, either by philosophers of science or historians of science, to the ramifications of such an approach for understanding science.[1] Philosophers of science, by and large, have imagined that they can lay bare the rationality of science by ignoring, in their analyses, the fact that scientific theories are usually attempts to solve specific empirical problems about the natural world.[2] Similarly, historians of science, for their part, have usually imagined that the chronology of scientific theories possesses an intrinsic intelligibility which requires little or no cognizance of the particular problems which prominent theories in the past were designed to solve.

It is the purpose of this short book to sketch what seem to be the implications, for both the history of science and its philosophy, of a view of scientific inquiry which perceives science as being—above all else—a problem-solving activity.

The approach taken here is not meant to imply that science is "nothing but" a problem-solving activity. Science has as wide a variety of aims as individual scientists have a multitude of motivations: science aims to explain and control the natural world; scientists seek (among other things) truth, influence, social utility, and prestige. Each of these goals could be (and has been) used to provide a framework within which one might try to explain the development and nature of science. My approach, however, contends that a view of science as a problem-solving system holds out more hope of capturing what is most characteristic about science than any alternative framework has.

As it becomes clear that many of the classic problems of philosophy of science, and many of the standard issues of the history of science, take on a very different perspective when we look at science as a problem-solving and problem-oriented activity, it will be argued that an attentive analysis of science from this perspective generates new insights which run counter to much of the "conventional wisdom" which historians and philosophers of science have taken for granted.

There is nothing modest about the claims this study makes. In brief, I shall be suggesting that a sophisticated theory of science qua problem-solving activity *must* alter the way we perceive both the central issues in the historiography of science and the central problems in the philosophy or methodology of science. I shall argue that if we take seriously the doctrine that the aim of science (and of all intellectual inquiry, for that matter) is the resolution or clarification of problems, then we shall have a very different picture of the historical evolution and the cognitive evaluation of science.

Before I contrast the problem-solving view of science with certain better known philosophies and histories of science, I must indicate specifically what I mean by a "problem-oriented theory of science." It is this preliminary goal which this chapter and the next aim to achieve.

The Nature of Scientific Problems

Throughout this essay, I shall be talking about what I call *scientific problems.* I should stress at the outset that I do not believe that "scientific" problems are fundamentally different from other kinds of problems (though they often are different in degree). Indeed, I shall show in chapter six that the view I am espousing can be applied, with only a few qualifications, to *all* intellectual disciplines. But, if we wish to study problem solving, we ought to begin with its most successful instances; so I shall limit my remarks in these preliminary sections largely to science itself.

If problems are the focal point of scientific thought, theories are its end result. Theories matter, they are *cognitively* important, insofar as—and only insofar as—they provide adequate solutions to problems. If problems constitute the questions of science, it is theories which constitute the answers. The function of a theory is to resolve ambiguity, to reduce irregularity to uniformity, to show that what happens is somehow intelligible and predictable; it is this complex of functions to which I refer when I speak of theories as solutions to problems.

Thesis 1: ***The first and essential acid test for any theory is whether it provides acceptable answers to interesting questions: whether,*** in other words, ***it provides satisfactory solutions to important problems.***

At one level, this might appear completely noncontroversial. Most writers who have dealt with the nature of science would probably claim to subscribe to such a view. Unfortunately, as we shall see, most philosophies of science manifestly fail to go so far as to justify even that seemingly harmless and obvious sentiment, let alone to explore its many ramifications.

The literature of the methodology of science offers us neither a taxonomy of the types of scientific problems, nor any acceptable method of grading their relative importance. It is noticeably silent about what the criteria are for an adequate solution to a problem. It does not recognize there are degrees of adequacy in problem solution, some solutions being better and richer than others. Insofar as contemporary philosophy of science says anything at all about these matters, it tends to regard all solutions on a par, and to assign all problems equal weight. In assessing the adequacy of any theory, the philosopher of science will usually ask how many facts confirm it, not how important those facts are. He will ask how many problems the theory solves, not about the significance of those problems. To this extent, contemporary philosophy of science has not captured the sense of thesis (1) above. It is for reasons such as these that I propose:

Thesis 2: *In appraising the merits of theories, it is more important to ask whether they constitute adequate solutions to significant problems than it is to ask whether they are "true," "corroborated," "well-confirmed" or otherwise justifiable within the framework of contemporary epistemology.*

But if it is plausible to think that the counterpoint between challenging problems and adequate theories is the basic dialectic of science, we must get a great deal clearer than we now are about what problems are and how they work, about how problems are weighted, and about the nature of theories and their precise relation to the problems which generate them (and which, as we shall see, they sometimes generate).

Empirical Problems

There are two very *different* kinds of problems which scientific theories are designed to solve. For now, I want to focus on the first, more familiar and archetypal, sense of the concept, which I shall call an *empirical* problem. Empirical problems are easier to illustrate than to define. We observe that heavy bodies fall toward the earth with amazing regularity. To ask how and why they so fall is to pose such a problem. We observe that alcohol left standing in a glass soon disappears. To seek an explanation for that phenomenon is, again, to raise an empirical problem. We may observe that the offspring of plants and animals bear striking resemblances to their parents. To inquire into the mechanism of trait transmission is also to raise an empirical problem. More generally, anything about the natural world which strikes us as odd, or otherwise in need of explanation, constitutes an empirical problem.

In calling such inquiry situations "empirical" problems, I do not mean to suggest they are directly given by the world as veridical bits of unambiguous data. Both historical examples and recent philosophical analysis have made it clear that the world is always perceived through the "lenses" of some conceptual network or other and that such networks and the languages in which they are embedded may, for all we know, provide an ineliminable "tint" to what we perceive. More to the point, *problems* of all sorts (including empirical ones) *arise within a certain context of inquiry* and are partly defined by that context. Our theoretical presuppositions about the natural order tell us what to expect and what seems peculiar or "problematic" or questionable (in the literal sense of that term). Situations which pose problems within one inquiry context will not necessarily do so within others. Hence, whether something is regarded as an empirical problem will depend, in part, on the theories we possess.

Why, then, call them "empirical" problems at all? I do so because, even granting that they arise only in certain contexts of theoretical inquiry, even granting that their formulation will be influenced by our theoretical commitments, it is nonetheless the case that we *treat* empirical problems as if they were problems about the world. If we ask, "How fast do bodies fall near the earth?", we are assuming there are objects akin to our conceptions of body and earth which move towards one another according to some regular rule. That assumption, of course, is a theory-laden one, but we nonetheless assert it to be about the physical world. Empirical problems are thus *first order problems;* they are substantive questions about the objects which constitute the domain of any given science. Unlike other, higher order problems (to be discussed in chapter two), we judge the adequacy of solutions to empirical problems by studying the objects in the domain.

We have already noted that there is an apparent functional similarity between talk of problems and problem solving and the more familiar rhetoric about facts and the explanation of facts.

Given that similarity, one might be inclined to translate the claims I shall make about the nature and logic of problem solving into assertions about the logic of explanation. To do so, however, would be to misconstrue the enterprise, for problems are very different from "facts" (even "theory-laden facts") and solving a problem can not be reduced to "explaining a fact." Full discussion of the disanalogies must wait until later, but some of the discrepancies can be seen by considering a few of the differences between facts or states of affairs on the one hand, and empirical problems on the other.

Certain presumed states of affairs regarded as posing empirical problems are actually *counterfactual.* A problem need not accurately describe a real state of affairs to be a problem: all that is required is that it be *thought to be* an actual state of affairs by some agent. For instance, early members of the Royal Society of London, convinced by mariners' tales of the existence of sea serpents, regarded the properties and behavior of such serpents as an empirical problem to be solved. Medieval natural philosophers such as Oresme, took it to be the case that hot goat's blood could split diamonds and developed theories to explain this counterfactual empirical "occurrence."[3] Similarly, early nineteenth century biologists, convinced of the existence of spontaneous generation, took it to be an empirical problem to show how meat left in the sun could transmute into maggots or how stomach juices could turn into tapeworms. For centuries, medical theory sought to explain the "fact" that bloodletting cured certain diseases. If factuality were a necessary condition for something to count as an empirical problem, then such situations could not count as problems. So long as we insist that theories are designed only to explain "facts" (i.e., true statements about the world), we shall find ourselves unable to explain most of the theoretical activity which has taken place in science.

There are many facts about the world which do not pose empirical problems simply because they are *unknown.* It is, for instance, presumably a fact (and always has been) that the sun is composed chiefly of hydrogen; but until the fact was discovered (or invented), it could not have generated a problem. In sum, a fact only becomes a problem when it is treated and recognized as such; facts, on the other hand, are facts, whether they are ever recognized. The only kind of facts which can possibly count as problems are *known* facts.

But even many known facts do not necessarily constitute empirical problems. To regard something as an empirical problem, we must feel that *there is a premium on solving it.* At any given moment in the history of science, many things will be well-known phenomena, but will not be felt to be in need of explanation or clarification. It was known since the earliest times, for instance, that most trees have green leaves. But that "fact" only became an "empirical problem" when someone decided it was sufficiently interesting and important to deserve explanation. Again, early societies knew certain drugs could produce hallucinations, but that widely known fact only became a recognized problem for physiological theories relatively recently.

Finally, problems recognized as such at one time can, for perfectly rational reasons, *cease* to be problems at later times. Facts could never undergo that sort of transformation. Early geological theorists, for instance, regarded one of the central problems of their discipline to be that of explaining how the earth took its shape within the last 6,000 to 8,000 years. With the elongation of the geological time scale, that staggering issue no longer remained a problem to be solved.

Types of Empirical Problems

Having seen some[4] of the differences between facts and empirical problems and the need for clearly separating the two, we can now turn to the role which such problems play in the process of scientific analysis. Although a fuller taxonomy will be developed later, we can roughly divide empirical problems into three types, relative to the function they have in theory evaluation: (1) *unsolved problems*—those empirical problems which have not yet been adequately solved by *any* theory; [5] (2) *solved problems*—those empirical problems which have been adequately solved by a theory; (3) *anomalous problems*—those empirical problems which a *particular* theory has not solved, but which one or more of its competitors have.[6]

Clearly, solved problems count in favor of a theory, anomalous ones constitute evidence *against* a theory, and unsolved ones simply indicate lines for future theoretical inquiry. Using this terminology, we can argue that *one of the hallmarks of scientific progress is the transformation of anomalous and unsolved empirical problems into solved ones.* Of any and every theory, we must ask how many problems it has solved and how many anomalies confront it. This question, in a slightly more complex form, becomes one of the primary tools for the comparative evaluation of scientific theories.

The Status of Unsolved Problems

It is part of the conventional wisdom that unsolved problems provide the stimulus for scientific growth and progress; and there can be no doubt that transforming unsolved into solved problems is one (though by no means the only) way in which progressive theories establish their scientific credentials. But it is too often assumed that the body of unsolved problems at any given time is clear cut and well defined, that scientists have a definite sense of which unsolved problems should be solved by their theories, and that a theory's failure to digest its unsolved problems is a clear liability.

A careful examination of many historical cases reveals, however, that the status of unsolved problems is a great deal more ambiguous than is often imagined. Whether a given "phenomenon" is a genuine problem, how important it is, how heavily it counts against a theory if it fails to solve it; these are all very complex questions, but a good first approximation to an answer is this: *unsolved problems generally count as genuine problems only when they are no longer unsolved.* Until solved by some theory in a domain they are generally only "potential" problems rather than actual ones.[7] There are two factors chiefly responsible for this: one, which we have already discussed, arises when we are unsure an empirical effect is genuine. Because many experimental results are difficult to reproduce, because physical systems are impossible to isolate, because measuring instruments are often unreliable, because the theory of error even leads us to expect "freak" results, it often takes a considerable time before a phenomenon is sufficiently authenticated to be taken seriously as a well-established effect. Second, it is often the case that even when an effect has been well authenticated, *it is very unclear to which domain of science it belongs* and, therefore, which theories should seek, or be expected, to solve it. Is the fact that the moon seems larger near the horizon a problem for astronomical theories, for optical theories, or for psychological ones? Is the formation of crystals and crystalline growth a problem for chemistry or biology or geology? Are "shooting stars" a problem for astronomy or for upper-atmosphere physics? Is the twitching of an electrified frog leg a problem for biology, chemistry, or electrical theory? We now have answers to all these questions and feel confident about assigning these problems to one domain or another. The chief reason for our confidence is that we have *solved* these problems. But for long periods in the history of science, these problems were unsolved and it was very unclear within what domain they should fall. As a result of that uncertainty, it did not count seriously against any theory in a given domain if it failed to solve these unsolved problems; for no one could show convincingly that theories in any particular domain should be expected to solve such problems.

The ambiguous status of unsolved problems is persuasively illustrated by the history of the problem of Brownian motion. First discussed at length by Robert Brown in 1828, it took the greater part of a century before scientists could decide whether it was a genuine problem, how important it was, and what sorts of theories should be expected to solve it. During the 1830s and 1840s for instance, it was alternately viewed as a biological problem (the Brownian particles perhaps being small "animalcules"), as a chemical problem, as a problem in the optics of polarization (e.g., by Brewster), as a problem of electrical conductivity (e.g., by Brongniart), as a problem in heat theory (e.g., by Dujardin), as a completely uninteresting mechanical effect which was too complicated and too insignificant to be worthy of efforts at solution, and—by some—as a nonproblem altogether.[8] So long as the problem remained unsolved, any theorist could conveniently choose to ignore it simply by saying that it was not a problem which theories in *his* field had to address. Yet this selfsame phenonemon, which could find neither home nor solution in the first half of the nineteenth century, gradually emerged as one of the core anomalies for classical thermodynamics and became, at the hands of Einstein and Perrin (who solved the problem), one of the triumphal successes of the kinetic-molecular theory of heat.

Consider, as another example, the famous case of Abraham Trembley's hydra-like polyp, first carefully observed in 1740. It was a phenomenon which seemed to run counter to the dominant biological ideas of the age; it could reproduce itself without sexual coupling and, when cut up, each part would quickly grow into a whole organism. These properties had been commonly observed in plants, yet were specifically denied to animals, suggesting that the polyp was a plant. On the other hand, the polyp had powers of locomotion, a stomach and patterns of food consumption usually associated with animals, especially insects. Here, then, was a living organism—half-plant, half-animal—whose very existence denied the long-cherished biological principle of three separable kingdoms (animal, vegetable, and mineral). The reaction to Trembley's discovery was immediate—throughout the 1740s and 1750s, biologists and naturalists all over Europe speculated about it and studied its behavior. This case would seem to be a compelling example of the generation of a serious empirical problem *in the absence of any theory which could solve it.*

But as Vartanian has convincingly shown,[9] the above account, suggesting as it does the emergence of an acute anomaly in the absence of any theoretical competition, is deplorably incomplete. What it ignores is the fact that—alongside of the dominant vitalistic biology—there existed a minority of biologists committed to a more materialistic and more mechanistic approach to biological processes. The regenerative powers of the polyp (along with its obvious animal characteristics) suggested that perhaps the materialists were correct. After all, if every part of the polyp, no matter how small, could regenerate a fully developed animal, then the materialists seemed to be right in denying there was an indivisible, super-materialistic soul which belonged to the whole organism only as an organized being.

Practically from the first discovery of the polyp, proponents of vitalistic biology recognized that the properties of the hydra could give "aid and comfort" to a rival research school. Cramer, Lyonnet, and two anonymous writers (in the *Mémoires de l'Académie des Sciences* and in the *Journal de Trévaux)* had, through the early and mid-1740s, already remarked on the susceptibility of the polyp to a materialistic interpretation (an interpretation fully developed by La Mettrie in his *L'homme machine).* In short, what transformed the polyp from an idle curiosity into a threatening anomaly for vitalistic biology was the presence of an alternative theory (or, as I shall later call it, an alternative research tradition) which could count the polyp as a solved problem.[10]

Cases in which there is doubt about the appropriate domain for some unsolved problem have frequently been of decisive historical importance. The vicissitudes of comets provide a neat example. During antiquity and the Middle Ages, comets were classified as sublunary phenomena and thus fell within the domain of meteorology. Astronomers, whose concern was exclusively with problems in the celestial regions, felt no need to offer theories about comets, nor even to plot their courses. By the sixteenth century, however, it had become customary to classify comets as celestial phenomena. This domain transition was crucial for the Copernican theory, since the motion of comets came to constitute one of the decisive anomalies for geocentric astronomy and one of the solved problems for the heliocentric theory.

One ought not conclude from their ambiguity that unsolved problems are unimportant for science, for transforming unsolved problems into solved ones is one of the means by which theories make empirical progress. But it must be stressed at the same time that a theory's failure to solve some unsolved problem generally will not weigh heavily against that theory, because we usually cannot know *a priori* that the problem in question should be soluble by that sort of theory. *The only reliable guide to the problems relevant to a particular theory is an examination of the problems which predecessor—and competing—theories in that domain (including the theory itself) have already solved.* Hence, in appraising the relative merits of theories, the class of unsolved problems is altogether irrelevant. What matters for purposes of theory evaluation are only those problems which have been solved, not necessarily by the theory in question, but by *some* known theory. (Here, as elsewhere, the evaluation of a theory is closely linked to a knowledge of its competitors.)

The Nature of Solved Problems

We have already indicated that "the solving of problems" ought not be confused with "the explaining of facts," and have discussed at some length the disanalogies between facts and empirical problems. What requires further elaboration is the difference between the logic and pragmatics of problem solution, and the logic and pragmatics of scientific explanation.

Most of the major differences emerge clearly if we begin by exploring the criteria for something to count as a solved problem. In very rough form, we can say that an empirical problem is solved when, within a particular context of inquiry, scientists properly no longer regard it as an unanswered question, i.e., when they believe they understand why the situation propounded by the problem is the way it is. Now clearly, it is theories which are meant to provide such understanding and any reference to a solved problem presupposes the existence of a theory which purportedly solves the problem in question. So when we ask whether a problem has been solved, we are really asking whether it stands in a certain relationship to some theory or other.

What does that relationship amount to? If we ask a logician of science the analogous question (to wit: what is the relation between an explanans and its explanandum?), he will generally tell us: the explaining theory must (along with certain initial conditions) entail an *exact* statement of fact to be explained; the theory must be either *true* or highly *probable;* whatever counts as an adequate explanation of any fact must be regarded as always having been such (so long as the epistemic appraisal of the explanans does not change). By way of contrast, I shall claim that: a theory may solve a problem so long as it entails even an *approximate* statement of the problem; in determining if a theory solves a problem, *it is irrelevant whether the theory is true or false, well or poorly confirmed;* what counts as a solution to a problem at one time will not necessarily be regarded as such at all times. Each of these differences demands further clarification.

The approximative character of problem solution. Although rare, it sometimes happens that a theory exactly predicts an experimental outcome. When that desirable result is achieved, there is cause for general rejoicing. It is far more common for the predictions deduced from a theory to come close to reproducing the data which constitute a specific problem, but with no exact coincidence of results. Newton was not able to explain exactly the motion of the planets; Einstein's theory did not exactly entail Eddington's telescopic observations; modern chemical bonding theory does not predict with exactitude the orbital distance of electrons in a molecule; thermodynamics does not precisely fit heat transfer data for any known steam engine. There are many reasons one could suggest (e.g., the use of "ideal cases," the non-isolation of real systems, imperfections in our measuring instruments) to explain the frequent small discrepancies between "theoretical results" and "laboratory results," but they are not of primary concern here. What is relevant is that facts are very rarely if ever explained (if we take our sense of explanation from the classical deductive model), because there is usually a discordance between what a theory entails and our laboratory data. By contrast, empirical problems are frequently solved because for problem solving purposes we do not require an exact, but only an approximate, resemblance between theoretical results and experimental ones. Newton did solve, and was widely regarded as having solved, the problem of the curvature of the earth—even though his results were not identical with observational findings. The thermodynamic theories of Carnot and Clausius were correctly perceived in the nineteenth century as adequate solutions to various problems of heat transfer, in spite of the fact that they applied exactly only to ideal (i.e., nonexistent) heat engines.

As should be clear, the notion of solution is highly relative and comparative in a way that the notion of explanation is not. We can have two different theories which solve the same problem, and yet say one is a better solution (i.e., a closer approximation) than the other. Comparable locutions and comparisons within the rhetoric of explanation are disallowed by many philosophers of science; on the standard model of explanation, something either is or definitely is not an explanation—degrees of explanatory adequacy are not countenanced. For instance, philosophers of science have been very troubled by the relationship of Galileo's and Newton's theories of fall and the data. Unable to say that both theories "explained" the phenomena of fall (because the two are formally inconsistent), they invented a variety of devices for excluding the title "explanatory" from one or the other of the theories. Yet it is surely more natural historically, and more sensible conceptually, to say that *both* theories (Galileo's and Newton's) solved the problem of free fall, one perhaps with more precision than the other (although even that is dubious). It redounds to the credit of both that, as Newton himself perceived, each provided an adequate solution to the problem at hand. We are, however, precluded from taking this natural way of describing the situation if we accept many of the current doctrines about the nature of explanation.

The irrelevance of truth and the falsity to solving a problem. The suggestion that questions of truth and probability are irrelevant when determining whether a theory solves a particular problem probably seems heretical, if only because one is so conditioned to considering the search for true understanding as one of the core aims of science. But whatever role questions of truth have in the scientific enterprise (and this is a large question to which we shall return[11]), one need not, and scientists generally do not, consider matters of truth and falsity when determining whether a theory does or does not solve a particular empirical problem.

We can all agree, for instance, that Ptolemy's theory of epicycles solved the problem of retrograde motion of the planets, regardless of whether we accept the truth of epicyclic astronomy. Equally, everyone agrees that Thomas Young's wave theory of light—whether true or false—solved the problem of the dispersion of light. Lavoisier's theory of oxidation, whatever its truth status, solved the problem of why iron is heavier after being heated than before. Generally, *any theory, T, can be regarded as having solved an empirical problem, if T functions (significantly) in any schema of inference whose conclusion is a statement of the problem.*

The frequent nonpermanence of solutions. One of the richest and healthiest dimensions of science is the growth through time of the standards it demands for something to count as a solution to a problem. What one generation of scientists will accept as a perfectly adequate solution will often be viewed by the next generation as a hopelessly inadequate one. The history of science is replete with cases where solutions whose precision and specificity were perfectly adequate for one epoch are totally inadequate for another. Consider a few examples:

In his *Physics,* Aristotle cites the problem of fall as a central phenomenon for any theory of terrestrial mechanics. Aristotle himself sought to understand both why bodies fall downwards and why they accelerate in fall. Aristotelian physics provided answers to these questions which were taken seriously for over two millenia. For Galileo, Descartes, Huygens, and Newton, however, Aristotle's views were not really solutions to the problem of fall at all, for they failed utterly to explain the "uniform difform" (i.e., uniformly accelerated) character of the fall of a body. One might want to say that the later thinkers were simply working with a very different problem than Aristotle was; I would be more inclined to see this as a case where, through the course of time, the criteria for what counts as solving a problem have evolved so much that what was once regarded as an adequate solution ceases to be regarded as such.

A clearer case is provided by the history of the kinetic theory of gases. By the 1740s, both Newton (using a central forces model) and Daniell Bernoulli (using a collision model) had shown that one could solve the problem of pressure-volume relations of gases in terms of assumptions about the mechanical interaction of their constituent particles. By the late nineteenth century, however, enough data about gaseous behavior had been accumulated to show that the simple kinetic theory provided only very inexact approximations to gaseous behavior, especially at low temperatures or high pressures. In short, given eighteenth-century standards of experimental accuracy and canons for adequacy of problem solution, the kinetic theory was a far cry from an adequate problem solver, especially so far as certain ranges of data were concerned. Accordingly, van der Waals and others set out to modify traditional kinetic theory so as to enable it to solve the problem of pressure-volume relations to meet their contemporary standards of problem solution. What resulted, of course, was van der Waal's equation.

In the history of many disciplines, humanistic as well as scientific, one can perceive a gradual tightening and strengthening of the threshold at which a theory will be conceded to be a solution to the relevant problem. Unless we acknowledge that the criteria for acceptable problem solutions do themselves evolve through time, the history of thought will seem enigmatic indeed.

The Special Role of Anomalous Problems

Many historians and philosophers of science have attached special significance to the place of anomalies in science. Thinkers from Bacon through Mill, Popper, Grünbaum and Lakatos have stressed the importance of refuting or falsifying experiments in the appraisal of scientific theories. Indeed, certain philosophies of science (especially those of Bacon and and Popper) make the search for, and resolution of, anomalies the *raison d'être* of the scientific enterprise, and the absence of anomalies the hallmark of scientific virtue. While sharing the view that anomalous instances have been, and should be, among the most important components of scientific rationality, I find myself seriously at odds with the conventional wisdom about what anomalies are and about the interpretation of their undoubted significance.

On the traditional view, anomalies have two chief characteristics:

(a) the occurrence of even one anomaly for a theory should force the rational scientist to abandon it;

(b) the only empirical data which can count as anomalies are those which are *logically inconsistent* with the theory for which they are anomalies.

I find these characteristics factually misleading as to actual scientific practice (both past and present), and a conceptual hindrance in understanding the role of anomalies in theory appraisal. I want to claim, by contrast, that:

(a′) the occurence of an anomaly raises doubts about, but need *not* compel the abandonment of, the theory exhibiting the anomaly;

(b′) anomalies need *not* be inconsistent with the theories to which they are anomalies.

The first of these two contentions (a′) is the less controversial, if only because numerous critics of the classical view have already offered cogent arguments for it; as a result, here I shall only briefly rehearse the reasons for it. The second thesis (b′), however, is not a familiar one, and I shall elaborate on it at some length.

Taking (a′) first, several philosophers (particularly Pierre Duhem, Otto Neurath, and W. Quine)[12] have argued that we cannot rationally decide whether a particular theory which generates an anomaly should be abandoned because of certain ineliminable *ambiguities* about the testing situation. The principal ambiguities are two:

1. In any empirical test, it is *an entire network of theories* which is required for deriving any experimental prediction. If the prediction turns out to be erroneous, we do not know where to locate the error within the network. The decision that one particular theory within the network is false is, these critics argue, completely arbitrary.

2. To abandon a theory because it is incompatible with the data assumes that our knowledge of the data is infallible and veridical. Once we realize that the data themselves are only probable, the occurrence of an anomaly does not necessarily require the abandonment of a theory (we might rationally choose, for instance, to "abandon" the data).

Still other critics[13] of (a) have stressed not the ambiguity, but the *pragmatics,* of theory testing and theory choice. They point out that almost every theory in history has had some anomalities or refuting instances; indeed, no one has ever been able to point to a single major theory which did not exhibit some anomalies. Accordingly, if we were to take (a) seriously, then we should find ourselves abandoning our entire theoretical repertoire in wholesale fashion, and thereby totally unable to say anything whatever about most domains of nature. For these reasons, there seem strong grounds for replacing (a) by the weaker, but more realistic, view (a′).

However, almost all writers on the subject of anomalies, whether defenders or critics of the classical view (a), seem to subscribe to (b), and to hold that an anomaly is only generated when there is a *logical* inconsistency between our "theoretical" predictions and our "experimental" observations. They have argued, in other words, that the only time data can be epistemically threatening for a theory is when such data contradict the claims of the theory. This strikes me as a far too restrictive notion of an anomalous problem. It is true, of course, that a genuine inconsistency between theory and observation may, under certain circumstances, constitute a particularly vivid example of anomaly. But such inconsistencies are far from being the only form of anomalous problem.

If, as I think we must, we take (a′) seriously, it is reasonable to characterize an anomaly as an empirical situation which, while perhaps not offering definitive grounds for abandoning a theory, does raise rational doubts about the empirical credentials of the theory. Proponents of (a′), in criticizing (a), are not claiming that we should ignore anomalies; rather, they are simply stressing that *anomalies constitute important, but not necessarily decisive, objections to any theory which exhibits them.* If we regard anomalies in this light (i.e., *as empirical problems which raise reasonable doubts about the empirical adequacy of a theory),* then we should abandon (b) for (b′), since by parity of reasoning, there are many empirical problems which, although consistent with a theory, can cast doubt upon its empirical foundations. Putting the point another way, there are occasions when scientists have *rationally* treated certain problems (which were consistent with a theory) in the same way that they would treat anomalies which were clearly inconsistent with the theory. Such situations arise when a theory in some field or domain fails to say anything about a kind of problem which other theories in the same domain have already solved.

Whether we treat such cases as anomalous depends, in part, of course, on our views about the aims of science. If one takes the narrow view that the object of science is simply to avoid making mistakes (i.e., false statements), then unsolved problems will not necessarily count seriously against a theory. But if one takes the broader view that science aims to maximize its problem-solving capacity (or, in more conventional language, its "explanatory content") then the failure of a theory to solve some well-recognized problem, which has been solved by a competitor theory, is a very serious mark against it. Ironically, most philosophers of science have paid lip service to the broader view, yet they have refused to recognize what that view entails—*the existence of a class of nonrefuting anomalies.*[14]

A careful look at the history of science makes it clear that a number of situations generate behavior similar to the kind of response which we have been led to expect when an inconsistency between theory and observation arises. *One of the most important species of anomaly arises when a theory, although not inconsistent with observational results, is nonetheless incapable of explaining or solving those results (which have been solved by a competitor theory).*[15] Thus, in Galileo's classic study of pendular motion, he criticizes the kinematical theories of his predecessors because they *cannot explain* the mathematics of pendular motion. His point is not that these earlier theories give an *incorrect* prediction for the geometry of the moving weight; rather his quarrel is that they give *no* prediction at all. Similarly, many critics of Newtonian celestial mechanics in the early eighteenth century argued that Newton's system of the world offered no explanation for the fact that all the planets move in the same direction around the sun, a phenomenon which had been solved by numerous previous astronomical theories, particularly Keplerian and Cartesian astronomy. Again, it is not that Newton's theory makes a *false* prediction about the direction of planetary revolution; rather, the flaw is that Newton's theory fails to address itself to the problem altogether. (It would be compatible with the Newtonian system, for instance, if adjacent planets moved in opposite directions.)

We can define this sort of anomaly more precisely by using some of the terminology set out above: *Whenever an empirical problem, p, has been solved by any theory, then p thereafter constitutes an anomaly for every theory in the relevant domain which does not also solve p.* Hence, the fact that some theory is logically consistent with *p* does not render *p* nonanomalous for that theory, if *p* has been solved by any other known theory in the domain.

The proposal, then, is that we should broaden our concept of an anomalous instance so as to include this important class of phenomena. Equally, in the spirit of (a′), we must weaken the epistemic threat of *all* anomalous instances by recognizing that, although anomalies constitute good grounds for arguing against a theory, they rarely, if ever, constitute final and decisive arguments against a theory. They are important to the delicate process of theory appraisal, but they remain but one of the vectors which determine the scientific acceptability of a theory.

In stressing that a problem can only count as *anomalous* for one theory if it is *solved* by another, the analysis seems to run against the common view that one sort of anomaly, *the refuting instance, poses a direct cognitive threat to a theory,* even if it is unsolved by any competitor. If a theory predicts a certain experimental outcome (say *O)* and experiment reveals that ∼*O* is the case, then surely, ∼*O* constitutes an anomaly for the theory even if no other theory can solve ∼*O?* As paradoxical as it may seem, this is generally unsound. An account of the reasons why many refuting instances are not anomalous requires further analytic machinery which will be developed in chapter three. We must here satisfy ourselves with the observation that unsolved refuting instances are often of little cognitive significance.

Converting Anomalies to Solved Problems

One of the most cognitively significant activities in which any scientist can engage is the successful transformation of a presumed empirical anomaly for a theory into a confirming instance for that theory. Unlike the solution of some new problem, the conversion of anomalies into problem-solving successes does double service: it not only exhibits the problem solving capacities of a theory (which the solution of any problem will do) but it simultaneously eliminates one of the major cognitive liabilities confronting the theory. This process of converting anomalies (real or apparent) into solved problems is as old as science itself; the history of ancient astronomy is replete with examples of it. Indeed, the basic idea is encapsulated in the classic aphorism *exceptio probat regulam*—which originally meant that a rule or principle is tested by its ability to deal with its apparent exceptions. Although numerous examples of this conversion phenomenon could be cited, the best known is probably the evolution of Prout's hypothesis concerning atomic composition. It was Prout's view that all the elements were composed of hydrogen and, consequently, the atomic weights of all elements should be integral multiples of the weight of hydrogen. Shortly after the appearance of this doctrine in 1815, numerous chemists pointed to seeming exceptions or anomalies. Berzelius and others found that several elements had atomic weights incompatible with Prout's theory (e.g., weights of 103.5 for lead, 35.45 for chlorine, and 68.7 for barium). These results constituted very serious anomalies for Proutian chemists. By the beginning of the twentieth century, however, the discovery of isotopes and the refinement of techniques of isotopic separation enabled physical chemists to separate out the isotopes of the same element; each isotope was found to have an atomic weight which was an integral multiple of hydrogen. The previously anomalous results could now be explained on Prout's hypothesis by showing them to be isotopic mixtures. Thus, *the very phenomena which had earlier constituted anomalies for Prout's hypothesis became positive instances for it.* Almost every major theory in the history of science has been able to produce comparable successes at digesting some of its initial anomalies.

The Weighting of Empirical Problems

Up to this point in the discussion, we have been assuming that all empirical problems are on essentially the same footing. In fact, of course, some solved problems count for more than others, and some anomalous problems are more threatening than others. If the problem-solving approach is ever to become a useful tool for appraisal, it must be able to show how, and why, certain problems are more significant than others.

The Weight of Solved Problems

There are certain empirical problems which, at a given time in a given scientific domain, are (and *should be)* given high priority; so high that if a theory in that domain solves them it will *ipso facto* be regarded as a serious contender for the rational allegiance of the scientific community. On the other hand, certain problems are of marginal importance. It would be nice to have a solution for them, but no theory is going to be abandoned simply because it fails to solve them. Similarly, anomalies range in importance from being decisive arguments against a theory (usually called "crucial experiments") to being rather minor exceptions which can often be completely ignored. If a philosophy of science, or a model of scientific progress, is going to be satisfactory, it must provide some guidelines not only for counting, but also for weighting, scientific problems on a scale of relative importance and cruciality.

In this section, I am going to be making some proposals concerning ways in which problems can be rationally weighted. Before embarking on that task, however, two caveats should be noted.

First, the criteria I am proposing are not meant to exhaust the modes of rational weighting. A calculus of problem weights is a major undertaking, well beyond the scope of this essay; hence, my list is only partial, suggestive rather than exhaustive.

Second, what follows concerns only the *cognitively rational weighting* of scientific problems. There are often occasions when a problem becomes of major importance to a community of scientists on nonrational or irrational grounds. Thus, certain problems may assume a high importance because the National Science Foundation will pay scientists to work on them or, as in the case of cancer research, because there are moral, social, and financial pressures which can "promote" such problems to a higher place than they perhaps cognitively deserve. It is not my purpose to discuss the nonrational dimensions of problem weighting (although I shall have something to say about that in chapter seven); we must first clarify what sorts of factors can affect the weighting of problems within the context of the rational appraisal of scientific theories.

In a new scientific domain, i.e., in a domain in which no adequate, systematic theories have yet been developed, almost all empirical problems are on a par. There is usually no good reason for singling out one, or a group of them, as being more important or crucial than another. Once we have one or more theories in the domain, however, we immediately have certain criteria for *increasing* the importance of certain empirical problems.[16] Three sorts of cases here are quite important:

Problem inflation by solution. If a particular problem has been solved by any viable theory in the domain, then that problem acquires considerable significance; to the extent that any competitor theory in the domain will almost certainly be expected either to solve it or to provide good grounds for failing to solve it. Thus, once Galileo found a solution to the problem of how fast bodies fall, every other subsequent theory of mechanics was under strong constraints to provide an equally adequate solution to the same problem.

Elaborating on an earlier point, it is tempting to formulate an even stronger version of this thesis by claiming that, in many (but not all) cases, an empirical situation does not even count as a problem at all until it has been solved by some theory in the domain. In such cases, solving a problem does not increase the previous weight of the problem; rather *it is the solution which allows us to recognize the problem as a genuine problem at all.* The reason for this is that it is often unclear whether a seeming problem really is an empirical problem, i.e., whether there is any natural phenomenon there to explain at all. Experiments in extrasensory perception are a case in point. Most scientists today would claim to be unsure that there is *any* evidence of ESP which is in need of theoretical explanation. The so-called "pseudo-sciences" (as well as newly emerging sciences) generally flourish on just such cases, where it is unclear whether there is, at the outset, any problem which needs to be solved.

Problem inflation by anomaly solution. If a problem has proved anomalous for, or resisted solution by, certain theories in the domain, then any theory which can transform that anomalous problem into a solved one will have strong arguments in its favor. The success of the special theory of relativity in solving the results of the Michelson-Morley experiments (which became anomalous problems for earlier aether theories) is a widely known example of just such a process. Other examples include: Newton's explanation of the shape of the earth and the elongation of the spectrum, Darwin's explanation of domestic breeding experiments, and Einstein's explanation of the photoelectric effect.

Problem inflation by archetype construction. At a more subtle level, there are other ways in which theories may endow certain empirical problems with greater significance than others. As we shall see later in detail, many theories single out, from the range of problems in the domain, certain empirical situations as archetypal. I call them "archetypal" because the theory indicates that they are the primary or basic natural process to which other processes in the domain must be reduced. For instance, before the time of Descartes, problems of the impact and collision of bodies were at the periphery of the concerns of writers on motion and mechanics, scarcely even recognized as problems which a theory of motion should resolve. But the mechanical philosophy of Descartes, precisely because it conceived collisions as the primary mode of interaction between bodies, promoted problems about impact to the forefront of mechanics, where they have remained ever since. In this case, as in other similar ones, the inflation of the value of collision problems was more than just a capricious shift of research emphasis. As a Cartesian, one was committed to the thesis that virtually the whole of natural science could be reduced to the laws of collision. But those laws, on which so much hung, were totally unknown early in the seventeenth century. It was thus entirely reasonable for Cartesians, and those interested in the Cartesian approach, to regard problems of impact and collision as among the most urgent in physics. Similarly, a century later Franklin's explanation of the Leyden jar, a primitive condensor, managed doubly to increase the significance of the problem of the Leyden jar, both by successfully solving what had already been recognized as a puzzling phenomenon, and by solving it utilizing a theory which made the Leyden jar an archetypical case of electrification, rather than the mere bizarre curiosity which it had generally been considered.[17]

What is notable about all three of the modes of problem weighting indicated above is the dependence of problem importance on the available theories. Without an appropriate type of theory, none of these three modes of problem weighting would be possible. There is, however, one type of problem weighting which is not always so dependent upon our existing theories:

Problem weighting by generality. There are sometimes occasions when one problem can be shown to be more general, and thus more important, than another. For instance, Kepler's problem of finding the law for the motion of Mars is presumably a special case of, and thereby less general than, his later problem of finding the law for the motion of all the planets. Mendel's problem of trait transmission in pea plants is clearly less general than the problem of trait transmission in all vegetables. But intuitions aside, the task of defining problem generality is a difficult one. A certain type of case is relatively straightforward: *if we can show for any two problems p' and p, that any solution to p' must also constitute a solution for p (but not vice versa) then p' is more general, and thus of greater weight, than p.* Although this represents an important class of cases, there are many other cases which do not permit one to evaluate their comparative generality. In such cases, we must fall back on the first three methods of differential weighting.

Just as such circumstances can render some problems more important than others, there are also circumstances which tend to *diminish* the importance of empirical problems, whether solved or unsolved.

Problem deflation by dissolution. As we have seen before, problems represent *presumed* states of affairs, assumptions about what we believe to be occurring in the world (or, more usually, in the laboratory). Because we sometimes change our beliefs about what is happening (if, for instance, certain experimental results cannot be reproduced), many problems simply vanish from a given domain. What was formerly regarded as an important problem may possibly cease to be one altogether, and becomes instead a "pseudo-problem." Even when the problem does not vanish entirely, its importance greatly diminishes as one's doubts about its authenticity or its relevance to the domain increase.

Problem deflation by domain modification. Another way in which the importance of a problem within a domain is signficantly diminished is by the expropriation of that problem by another domain. Until the early seventeenth century, for instance, writers on physical optics felt it important to explain what was known about the physiology of the eye and the psychology of vision. No "optical" theory was adequate unless it addressed these problems. With the increasing specialization of knowledge, however, problems in the physiology of vision and in the psychology of perception were excised out of physical optics, and thus had their prior importance within optics radically devalued.

Problem deflation by archetype modification. As we saw above, certain problems can be given prominence by the emergence of a new theory which gives them special importance. The mirror image of this process occurs when a theory is repudiated. Those problems which came into prominence because they were archetypes of a now abandoned theory may lose some of their importance as the theory with which they were so closely allied wanes. After, for instance, Descartes and other physicists in the seventeenth century had succeeded in making collision processes into the archetypal mechanical process, situations of work and energy expenditure—which had been among Aristotle's core examples—lost much of their earlier prominence.

The Weight of Anomalous Problems

It has often been maintained, particularly by Karl Popper but generally by all the logical empiricists, that any theory which had anomalous empirical problems (in their language, a theory which had been "refuted" or "disconfirmed") was no longer worthy of serious scientific consideration. Any anomaly, any "refuting instance," was as important as any other. And one empirical anomaly for a theory was as devastating as a hundred. It has recently become clear, however, that such an approach will not do; certainly not in practice and probably not even in principle. As Kuhn and others have stressed, virtually every theory ever devised, *including* those accepted by scientists today, has anomalous instances. It simply is not true that, in general, the discovery of an anomaly for a particular theory will lead, in and of itself, to the abandonment of the theory which exhibits the anomaly. At the same time, we must recognize that there have been circumstances when theories were confronted with sufficiently acute anomalous instances that they were abandoned. If we are to capture whatever modicum of rationality is implicit in such activity, we must be able, at least roughly, to grade the anomalies which confront a theory in order to indicate at least the differences between those anomalies which are disastrous for a theory and those which are only a mild embarrassment.

One possible way of dealing with this dilemma has been offered by Thomas Kuhn, who proposes essentially that it is the accumulation of *a large number of anomalies* which finally induces scientists to abandon a theory.[18] The difficulties with Kuhn's solution to this problem are manifold: Kuhn offers no reason why, for any number of anomalies, *n,* scientists should be undisturbed by *n-1* anomalies and suddenly ready to abandon the theory altogether when it has *n* anomalies; Kuhn's account cannot be squared with the historical fact that scientists have often abandoned a theory in the face of only a few anomalies and have other times retained a theory in the face of an ocean of empirical refutations.

I submit that if we are to find any rhyme or reason in the role of anomalies within the history of science it can come only by a recognition that it is not so much *how many* anomalies a theory generates that count, but rather *how cognitively important* those particular anomalies are.

How, then, can we begin to grade the importance of empirical anomalies? The most natural approach here would seem to involve grading anomalies in terms of *the degree of epistemic threat* which they pose for a theory. A preliminary first step in this direction comes from the recognition that the importance of any particular anomaly for a theory depends largely on the competitive state of play between that theory and its competitors. If a theory happens to be the only known theory in a particular domain, then it may have dozens of "refuting" instances and probably none of them will be of decisive importance. After all, when we ask about the importance of anomaly, we are really asking the question: To what degree should that anomaly dispose us to abandon the theory which generated it? If there is no alternative theory in view to displace it, all thought of abandoning it is probably academic, for, in the absence of a successor, that would be a cognitive defeat of the first order. So, *assessing the importance of any seemingly anomalous problem for a theory has to be done within the context of the other competing theories in the domain.* Given that such theories exist, we can then ask whether a particular unsolved problem, exhibited by T_1, is also exhibited by T_1's competitors. If the answer is affirmative, that is, if all the extant theories in the domain find themselves equally unable to solve that particular phenomenon, then that problem cannot loom very large in the assessment of T_1—even if the problem is logically inconsistent with T_1. If, on the other hand, there is some empirical problem which is unsolved by T_1, but for which some competing theory is able to provide a solution, then that unsolved problem assumes a considerable significance for T_1; in short it becomes a genuine anomaly. Clearly, the importance of an anomaly for a theory can vary enormously with time and circumstances.

An example or two may make this clearer. Scientists since antiquity have recognized that any astronomical and optical theory ought to be able to explain the color of the sky. Until the early twentieth century, however, no theory was able to provide any adequate explanation why light, passing through empty space and being refracted into the atmosphere, should produce the familar blue color. It was only after Rayleigh worked out a theory of atmospheric dispersion that the inability of an optical theory to explain the blue of the sky counted as a major argument against such a theory. Similarly, the capacity of friction to produce heat was a long known counterinstance to the view that heat was a substance which inhered in bodies. But it was only after the development of a kinetic theory of heat which could successfully deal with the generation of heat by friction that heat by friction became an important problem for heat-as-substance theories. But the discussion thus far only tells us how to identify an anomaly, not how to grade its importance.

One important determinant of the importance of an anomaly is *the degree of discrepancy* between the observed experimental result and the theoretical prediction. Every theory is constantly confronted by small order discrepancies between what it predicts and what is observed. In the absence of a theory which exhibits better fit with the data, few persons would attach much importance to such quasi-anomalies. More serious, however, are those discrepancies which are large, often representing several orders of magnitude. Scientists are prepared to live with theories which are approximate, but only to a certain degree. Precisely where to draw the line depends very largely on the conventional standards of accuracy, both theoretical and experimental, within the domain. It is clear, for instance, that cosmologists or geologists are often prepared to attach much less significance to seemingly large discrepancies between predicted and observed results than is, say, a physical chemist or a spectroscopist. These differences in precision tolerance in the various disciplines do not mean that these tolerance limits are arbitrary. To the contrary, they usually reflect the subtle instrumental and mathematical constraints on the field, as well as the complexity of the process under investigation. What is common to all the sciences is a conviction that certain experimental results are so discordant as to constitute acutely important anomalies, whereas other, only mildly disconcordant results are relatively minor problems. Here again, the comparative state of play between competing theories is decisive.

A second factor which influences the weight of an anomaly is its *age and its demonstrated resistance to solution by a particular theory.* No one is very surprised if a newly discovered phenomenon (perhaps anticipated predictively by one theory) is anomalous for some other theory in the domain. Experience teaches us that it sometimes takes a number of intra-theoretic adjustments before a problem can be convincingly solved. If, however, after repeated efforts, a theory remains unable to explain the anomaly, then it comes to loom ever larger as an epistemic embarrassment. It is for this reason, incidentally, that so-called crucial experiments—designed to choose between competing theories—are rarely decisive immediately. It takes a certain amount of time and effort at reconciliation before one can reasonably come to the conclusion that a theory is probably going to be unable to solve any given anomalous problem.

I shall have more to say later about this general issue of the weighting of empirical problems; but we can summarize the discussion thus far by stressing two central claims:

1. the importance of solving all empirical problems (whether solved or anomalous) is not the same, some being of much greater weight than others;
2. the assessment of the importance of a particular problem or anomaly requires a knowledge of the various theories within the domain *and* a knowledge of how successful or unsuccessful those theories have been at offering solutions.

Theory Complexes and Scientific Problems

Up to this point in the discussion, I have been writing as if it were *single* theories which solve, or fail to solve, empirical problems. I have argued that individual theories can take credit for the problems they solve and that they must bear the blame for those anomalies they generate. It might be said, however, that in taking such an approach I have ignored one of the most striking and most significant aspects of the testing situation; namely, the ambiguity of *the epistemic threat posed by anomalies*. In order to determine whether my analysis founders on this point, we must examine the arguments for the ambiguity with some care.

The Alleged Ambiguity of Theory Testing

In the early years of this century, the French physicist-philosopher Pierre Duhem argued that the testing of theories is a great deal more complicated than the uncritical observer might imagine.[19] He pointed out that individual theories do not usually entail anything that can be directly observed in the laboratory; rather, it is, he maintained, only a *complex conjunction* of a variety of theories which can ever lead (given certain statements of initial conditions) to any predictions about the world. For instance, in order to test a theoretical statement as simple as Boyle's law, we must invoke (among other things) theories about the behavior of our measuring instruments. Boyle's law by itself predicts nothing whatever about how those instruments will behave. If, then, it is always (or even usually) the case that theory complexes rather than individual theories are subjected to empirical test, certain crucial ambiguities seem to arise. Suppose, for instance, that a theory complex produces an erroneous result (i.e., it leads to a prediction which is refuted by the evidence). What conclusion can we draw from that? Duhem (and most of his recent commentators) wants to argue that we can never deduce with certainty which theoretical element(s) in the complex has been refuted or falsified by the recalcitrant observation. All we learn from experience, he says, is that we have gone astray somewhere, but the logic of scientific inference is too imprecise to allow us with certainty to pin the blame on any particular component or components in the theoretical complex. It follows that we can never legitimately claim that any theory has ever been refuted.[20]

A similar, but hitherto unnoted, ambiguity apparently affects the confirmation as well as the refutation of individual scientific theories of hypotheses. If it is true that theory complexes, and only theory complexes, can confront experience, then *the successful prediction of an experimental outcome leaves us in as much doubt about how to distribute credit, as an unsuccessful prediction leaves us unclear where to locate blame.* In the case of a successful confirmation, should we assume that each member of the theory complex is confirmed by the outcome? And should we assume that each member gets the same incremental increase in its degree of confirmation as every other member? These are difficult, and I think still unanswered, questions.

But what are we to make of these ambiguities of testing as far as the model being discussed here is concerned? Is that model open, in the ways that the received view is, to such an analysis and do these ambiguities make it meaningless to talk about the appraisal of individual theories and hypotheses?

Problem Solving and Ambiguous Tests

I shall show below that the ambiguities of testing, while genuine enough and worrying when directed against the standard mode of discussing theory appraisal, are relatively harmless when seen within the context of a problem-solving model of theory appraisal. I shall show, further, that—within the latter model—there is a natural way of handling the Duhemian ambiguities which will still allow us to talk of the rational appraisal of *individual* theories without having to retreat to talking exclusively of theory complexes.

Let us deal with the ambiguities of refutation or falsification first. The argument there, we recall, concluded that we cannot legitimately deduce the falsity of any component of a theory complex from the falsity of the theory complex as a whole. For the sake of discussion, let us grant that this argument is conclusive. Even if cogent, it implies *nothing whatever* about the appropriateness of appraising the problem-solving effectiveness of individual theories. We might, for instance, entirely consistently with Duhemian worries, adopt the following principle (A_1),

> Whenever any theory complex, *C,* encounters an anomalous problem, *a,* then *a* counts as an anomaly for *each nonanalytic element,* $T_1, T_2, \ldots, T_n$, of *C.*[21]

Why is the principle (A_1) immune from criticism of a Duhemian type? Simply because the whole thrust of the Duhemian analysis has to do with *assignments of truth or falsity* (or such weaker surrogates for them as probability or degree of confirmation) to individual theories. The cogency of Duhem's position (as well as its recent elaborations) depends upon the peculiar features of the assignment of truth-values within a *modus tollens* argument. In that argument schema, we are asked to imagine a situation where a theory complex, *C,* entails some observation, *O,* which is false:

[C(consisting of $T_1, T_2, \ldots, T_n$) + initial conditions] $\rightarrow$ O

Not-O is observed

The Duhemian points out that logic does not permit the assertion of the falsity of any element, T_i, of the complex just because the complex itself has been falsified.

Within the problem-solving model, however, we make no assignments of truth or falsity; there is nothing in the structure of deductive logic which precludes the localization of properties such as problem-solving effectiveness. When we say that *a* is an anomaly for a theory T_1, we are not saying that *a* falsifies T_1 (to claim that would open oneself to Duhemian objections); rather, we are saying that *a* is the sort of problem which a theory such as T_1 ought to be able to solve (albeit in conjunction with other theories), but which it has failed as yet to solve. That, of course, does *not* prove that T_1 is false; but it *does* clearly raise doubts about the problem-solving effectiveness of T_1 (and, for that matter, about *every other* T_i *in the complex* that failed to solve the empirical problem *a)*.

A similar sort of analysis applies to the apparent ambiguities of confirmation. When we stress those ambiguities, it is because we are not clear how much a successful confirmation of a theory complex ought to increase our confidence in the *truth* (or the likelihood) of its component elements. But if we shift from talk about truth or probability to talk about problem solving, this ambiguity dissolves as well, for there is a mirror image here of the principle (A_1) defined above for anomalies; namely (A_2),

> Whenever any theory complex, *C,* adequately solves an empirical problem, *b,* then *b* counts as a solved problem for *each* nonanalytic element, T_1, T_2, . . . , T_n of *C.*

As principles (A_1) and (A_2) make clear, I am proposing we turn the usual response to these Duhemian ambiguities on its head. Where previous writers on this issue have tended to imagine that the solution to the Duhemian ambiguity consists in trying to find some way, *contra* Duhem's analysis, for *localizing* blame or credit, I want to try the opposite approach by suggesting that a way out of the Duhemian conundrum may emerge if, far from *localizing* blame or credit in one place, we simply *spread it evenly among the members of the complex* (using a rational variant of the guilt-by-association doctrine).

A full argument for principles (A_1) and (A_2) requires lengthier treatment than I can give here. What I shall claim, however, is that there is nothing in the usual arguments for the ambiguity of testing which would undercut (A_1) or (A_2). To that extent at least, we are entitled to claim that it seems to be entirely appropriate to talk about the appraisal of individual theories—with the proviso that such appraisals concern problem-solving effectiveness and not truth or falsity.

There is yet another important dimension of the Duhemian problem which must be mentioned here, although a thorough treatment of it will have to wait until we have developed further machinery for theory appraisal in the next chapter. The dimension in question has to do with the nature of a rational response to a so-called falsifying experiment. On my analysis, whenever a complex of theories generates an anomaly, that anomaly counts against *each* element within the complex. The fact that each of those theories has this particular anomaly does not, of course, require that they should each be abandoned; for, as we have seen, the existence of an anomalous problem for a theory is not *ipso facto* sufficient grounds for abandoning the theory. But that is not an end on it. Precisely because the anomaly exists, and because science seeks to minimize anomalies, there is still cognitive pressure on the scientific community to attempt to resolve the anomaly. Resolving that anomaly will require, presumably, the abandonment (though not by virture of its "falsification") of at least one of the theories that composed the complex that was unable to deal with the anomaly. From my point of view (and I suspect that from Duhem's too), *the real challenge of the Duhemian analysis consists,* not in showing how we can "localize" falsehood or truth, but rather *in showing what rational strategies there are for selecting a better complex.*[22] It is this point to which I shall return in chapter three, where machinery for making the relevant assessments will be described.

Chapter Two
Conceptual Problems

If a historian accepts the [customary] analysis of confirmation, . . . he may conclude that the course of scientific development is massively influenced by . . . nonevidential considerations. WESLEY SALMON (1970), p. 80

Our discussion in chapter one focussed exclusively on empirical problems and on the connections between such problems and the theories which purport to solve them. It would be an enormous mistake, however, to imagine that scientific progress and rationality consist entirely of solving empirical problems. There is a second type of problem-solving activity which has been *at least as important* in the development of science as empirical problem solving. This latter type of problem, which I call a *conceptual problem,* has been largely ignored by historians and philosophers of science (though rarely by scientists), presumably because it does not comport well with those empiricist epistemologies of science which have been the reigning fashion for more than a century. The purpose of this chapter is to state the case for a richer theory of problem solving than empiricists have allowed, to explore the nature of these nonempirical problems and to show what role they have in theory appraisal.

Even the briefest glance at the history of science makes it clear that the key debates between scientists have centered as much on nonempirical issues as on empirical ones. When, for instance, the epicyclic astronomy of Ptolemy was criticized (as it often was in antiquity, the Middle Ages and the Renaissance), the core criticisms did *not* deal with its adequacy to solve the chief empirical problems of observational astronomy. It was readily granted by most of Ptolemy's critics that his system was perfectly adequate for "saving the phenomena." Rather, the bulk of the criticism was directed against the conceptual credentials of the mechanisms Ptolemy utilized (including equants and eccentrics, as well as epicycles) for solving the empirical problems of astronomy. Similarly, the later critics of Copernican astronomy did not generally claim it was empirically inadequate at predicting the motions of celestial bodies; indeed, it could solve some empirical problems (such as the motion of comets) far better than the available alternatives. What chiefly troubled Copernicus' critics were doubts about how heliocentric astronomy could be integrated within a broader framework of assumptions about the natural world—a framework which had been systematically and progressively articulated since antiquity. When, a century after Copernicus, Newton announced his "system of the world," it encountered almost universal applause for its capacity to solve many crucial empirical problems. What troubled many of Newton's contemporaries (including Locke, Berkeley, Huygens, and Leibniz) were several conceptual ambiguities and confusions about its foundational assumptions. What was absolute space and why was it needed to do physics? How could bodies conceivably act on one another at-a-distance? What was the source of the new energy which, on Newton's theory, had to be continuously super-added to the world order? How, Leibniz would ask, could Newton's theory be reconciled with an intelligent deity who designed the world? In none of these cases was a critic pointing to an unsolved or anomalous empirical problem. They were, rather, raising acute difficulties of a *nonempirical kind.* Nor is it merely "early" science which exhibits this phenomenon.

If we look at the reception of Darwin's evolutionary biology, Freud's psychoanalytic theories, Skinner's behaviorism, or modern quantum mechanics, the same pattern repeats itself. Alongside of the rehearsal of empirical anomalies and solved empirical problems, both critics and proponents of a theory often invoke criteria of theoretical appraisal which have nothing whatever to do with a theory's capacity to solve the empirical problems of the relevant scientific domain.

Of course, this pattern has not gone unnoticed by historians, philosophers and sociologists of science; it is too obvious and too persistent to have been ignored altogether. But the usual response, when confronted with cases in which theories are being appraised along nonempirical vectors, has been to deplore the intrusion of these "unscientific" considerations and to attribute them largely to prejudice, superstition, or a "prescientific temperament." Some scholars (such as Kuhn) have gone so far as to make the absence of such nonempirical factors a token of the "maturity" of any specific science.[1] Rather than seeking to learn something about the complex nature of scientific rationality from such cases, philosophers (with regret) and sociologists (with delight) have generally taken them as tokens of the irrationality of science as actually practiced.[2] As a result few scholars who study the nature of science have found any room in their models for the role of such conceptual problems in the rational appraisal of scientific theories.[3] Empiricist philosophies of science (including those of Popper, Carnap and Reichenbach) and even less strident empiricist methodologies (including those of Lakatos, Collingwood and Feyerabend)—all of which imagine that theory choice in science should be governed exclusively by empirical considerations—simply fail to come to terms with the role of conceptual problems in science, and accordingly find themselves too impoverished to explain or reconstruct much of the actual course of science. Such empiricist theories of science exhibit particularly awkward limitations in explaining those historical situations in which the empirical problem-solving abilities of competing theories have been virtually *equivalent.* Cases of this kind are far more common in science than people generally realize. The debates between Copernican and Ptolemian astronomers (1540-1600), between Newtonians and Cartesians (1720-1750), between wave and particle optics (1810—1850), between atomists and anti-atomists (1815 to about 1880) are all examples of important scientific controversies where the empirical support for rival theories was essentially the same. Positivistically inspired accounts of these historical encounters have shed very little light on these important cases: this is scarcely surprising since the positivist holds empirical support to be the only legitimate arbiter of theoretical belief. These controversies must, by the strict empiricist, be viewed as mere *querelles de mots,* hollow and irrational debates about issues which experience cannot settle.

A broader view concerning the nature of problem solving—one which recognizes the existence of conceptual problems—puts us in a position to understand and to describe the kind of intellectual interaction that can take place between defenders of theories which are equally supported by the data. Because the assessment of theories is a multi-factorial affair, parity with respect to one factor in no way precludes a rational choice based on disparities at other levels.

The Nature of Conceptual Problems

Thus far, we have defined conceptual problems by exclusion, suggesting that they are nonempirical. Before we can understand their role in theory appraisal, we must clarify precisely what they are and how they arise. To begin with, we must stress that a conceptual problem is a problem *exhibited by some theory or other.* Conceptual problems are characteristics of theories and have no existence independent of the theories which exhibit them, not even that limited autonomy which empirical problems sometimes possess. If empirical problems are first order questions about the substantive entities in some domain, conceptual problems are higher order questions about the well-foundedness of the conceptual structures (e.g., theories) which have been devised to answer the first order questions. (In point of fact, there is a continuous shading of problems intermediate between straightforward empirical and conceptual problems; for heuristic reasons, however, I shall concentrate on the distant ends of the spectrum.)

Conceptual problems arise for a theory, T, in one of two ways:

1. When T exhibits certain internal inconsistencies, or when its basic categories of analysis are vague and unclear; these are *internal conceptual problems.*
2. When T is in conflict with another theory or doctrine, T', which proponents of T believe to be rationally well founded; these are *external conceptual problems.*

Each of these forms of conceptual problems needs to be analyzed in some detail.

Internal Conceptual Problems

The most vivid, though by no means the most frequent, type of internal conceptual problem arises with the discovery that a theory is logically inconsistent, and thus self-contradictory. Probably most common in the history of mathematics, inconsistent theories have often been detected in almost all the other branches of science.[4] Little need be said about the acuteness of such problems. Unless the proponents of such theories are prepared to abandon the rules of logical inference (which provided the groundwork for recognizing the inconsistency), or can somehow "localize" the inconsistency, the only conceivable response to a conceptual problem of this kind is to refuse to accept the offending theory until the inconsistency is removed.[5]

More common, as well as more difficult to handle, are a second class of internal conceptual problems; namely, *those arising from conceptual ambiguity or circularity within the theory.* Unlike inconsistency, the ambiguity of concepts is a matter of degree rather than kind. Some degree of ambiguity is probably ineliminable in any except the most vigorously axiomatized theories. It may even be true that some small measure of ambiguity is a positive bonus, since less rigorously defined theories can often be more readily applied to new domains of investigation than more rigid ones. But granting that, it is nonetheless true that systematic and chronic ambiguity or circularity within a theory often has been, and should be, viewed as highly disadvantageous.

Examples of such conceptual problems abound in the history of science. For instance, Faraday's early model of electrical interaction was designed to eliminate the concept of action-at-a-distance (itself a conceptual problem in earlier Newtonian physics). Unfortunately, as Robert Hare showed,[6] Faraday's own model required short range actions-at-a-distance. Faraday had merely replaced one otiose concept by its virtual equivalent. Even worse, Faraday's model—as Hare was quick to point out—postulated "contiguant" particles, which were not really contiguous at all. These kinds of criticisms led Faraday to re-think his views on matter and force and were eventually responsible for the emergence of Faraday's field theory, which avoided these conceptual problems. Taking another example from nineteenth-century physics, it was often alleged by the critics of the kinetic-molecular theory (e.g., Stallo and Mach) that the kinetic theory was nonexplanatory because circular. For instance, it explained the elasticity of gases by postulating elastic constituents (i.e., molecules). But, observed the critics, because we understand no more about the causes of elasticity in solids than we do in fluids, the kinetic explanation is entirely circular.[7]

The increase of the conceptual clarity of a theory through careful clarifications and specifications of meaning is, as William Whewell observed more than a century ago, one of the most important ways in which science progresses. He called this process "the explication of conceptions" and showed how a number of theories, in the course of their temporal careers, had become increasingly precise—largely as a result of the critics of such theories emphasizing their conceptual unclarities.[8] Many important scientific revolutions (e.g., the emergence of the theory of special relativity, the development of behavioristic psychology) have depended largely on the recognition, and subsequent reduction, of the terminological ambiguity of theories within a particular domain.

Although both these types of internal problems are doubtlessly important in the process of theory appraisal, neither have played as decisive a historical role as the other kinds of conceptual problems have.

External Conceptual Problems

External conceptual problems are generated by a theory, *T*, when *T* is in conflict with another theory or doctrine which the proponents of *T* believe to be rationally well founded. It is the existence of this "tension" which constitutes a conceptual problem. But what precisely do the "tension" and the "conflict" amount to? The easiest form of "tension" to define, although by no means the most frequent, is that of *logical inconsistency* or *incompatibility.* When one theory is logically inconsistent with another accepted theory, then we have a vivid example of a conceptual problem.

The development of astronomy in ancient Greece, to which we have already referred, provides a useful case in point. The unsolved empirical problem here (it was actually a host of related problems) was summarized in tables of planetary motion, tables which recorded the apparent positions of the sun, moon, and planets at different times. This was the initial empirical problem which had to be resolved. The succession of planetary theories in antiquity, from the homocentric spheres of Eudoxus and Aristotle to the complex epicycles, eccentrics, and equants of Ptolemy, illustrates a series of attempts to solve the problems of early astronomy. But as soon as the early astronomical theories were developed each of them in turn generated a plethora of other problems, some of them empirical, others conceptual. Thus, the homocentric spheres of Eudoxus and Aristotle failed to explain accurately the retrogradations of the planets and the seasonal inequalities exhibited by the data. These phenomena were clearly recognized as unsolved problems. On the other hand, the later system of Ptolemy managed to avoid most of the anomalous problems which earlier Greek astronomy had encountered, but the price it paid to do so was that of *generating enormous conceptual problems.* Ever since the time of Plato, astronomers had worked on the assumption that the heavenly motions were "perfect" (i.e., that each planet moved in a perfect circle about the earth at constant speed). This assumption put enormous constraints on the kinds of hypotheses which were open to astronomers. Ptolemy's system, for all its empirical virtues, ran afoul of these prohibitions by making assumptions about the behavior of celestial bodies (e.g., the hypothesis that certain planets move around empty points in space, that planets do not always move at constant speed, and the like) which were in flagrant contradiction with the then universally accepted physical and cosmological theories about the nature and motion of the heavenly bodies. In spite of ingenious efforts to reconcile these differences by Ptolemy and others, most of the crucial conceptual problems remained, and were to plague the development of mathematical astronomy until the end of the seventeenth century (and even beyond).

But there are other relations besides that of inconsistency which also constitute conceptual problems for those theories which exhibit them. One common situation arises when two theories, although logically compatible, are jointly *implausible,* i.e., when the acceptance of either one makes it less plausible that the other is acceptable. For example, many late seventeenth-century theories of physiology were based on the (Cartesian) assumption that the various bodily processes were essentially caused by the mechanical processes of collision, filtration, and fluid flow. Once Newtonian physics was accepted, many critics of mechanistic physiology pointed out that such mechanistic doctrines, although logically compatible with the physics of Newton, were nonetheless rendered rather implausible by Newtonian physics. The argument went something like this: Newtonian physics, while certainly allowing for the existence of collision phenomena, nonetheless shows that *most* physical processes depend upon more that the impacts between, and the motions of, particles. To the extent that "mechanistic" (Cartesian inspired) theories of physiology postulate such processes as the *exclusive* determinant of organic change, they rest on a huge improbability. They are consistent with Newtonian physics (for that physics does not deny that there can be some material systems which are entirely mechanical); but it did seem highly implausible, given Newtonian physics, that a system as complex as a living organism could function with only a limited range of the processes exhibited in the inorganic realm.

A second example may clarify the notion of conceptual problem-generation by joint implausibility between theories. Throughout the seventeenth and early eighteenth century, the dominant theory of heat was a *kinetic* one; heat was conceived as the rapid agitation of the constituent parts of a body. Throughout the eighteenth century, however, a number of theories in a variety of fields began to suggest that many natural processes depended upon the presence of one or more highly elastic, highly rarefied fluids which could be absorbed by, or released from, material bodies. Although electricity was the best known example, such subtle fluids were postulated to explain magnetism, neurological functioning, perception, embryology, and even gravity. As these theories became more widely accepted, and as certain observable analogies between heat, light and electricity began to be explored, kinetic theories of heat came under sustained attack. While the acceptance of, for example, a fluid theory of electricity did not entail the denial of the kinetic theory of heat, it was thought that kinetic theories of heat became increasingly implausible as one domain after another came to be dominated by highly successful ideas about the substantial, as opposed to the kinetic, nature of physical processes.

A third manner in which conceptual problems can be generated occurs when a theory emerges which ought to reinforce another theory, but fails to do so and is *merely compatible with it.* To understand what is involved in such cases, we must talk briefly about the interdisciplinary structure of science, for compatibility between two systems or theories is not, in common parlance, regarded as a sign of cognitive weakness. The various scientific disciplines and domains are never completely independent of one another. At any given epoch, there are hierarchical systems of interconnection between the various sciences which condition the rational expectations which scientists have when they appraise theories. In our own time, for instance, it is presumed that the chemist will look to the physicist for ideas about atomic structure; that the biologist should utilize chemical concepts when talking about organic microstructures. The enunciation of a chemical theory which was merely compatible with quantum mechanics, but which utilized none of the concepts of quantum theory, would be viewed askance by most modern scientists. Similarly, a theory of heredity which was compatible with chemistry but failed to exploit any of its analytic machinery, would likewise be suspect. Different epochs, of course, will have different expectations about which disciplines should borrow from, and reinforce, others. (In the seventeenth century, for instance, it was expected that any physical theory should be positively relevant to, and not merely compatible with, Christian theology.)

As should be clear, *mere* compatibility between two theories is not always a conceptual problem. No one thinks, for instance, that a theory in micro-economics is flawed if it is merely compatible with thermodynamics. But in many cases, compatibility, as opposed to positive relevance, between two theories is quite rightly viewed as a major drawback to the acceptance of the theories in question.

Our discussion thus far puts us in a position to outline a taxonomy of the various cognitive relationships which can exist between two (or more) theories:

1. *Entailment*—one theory, T, entails another theory, T_1.
2. *Reinforcement*—T provides a "rationale" for (a part of) T_1.[9]
3. *Compatibility*—T entails nothing about T_1.
4. *Implausibility*—T entails that (a part of) T_1 is unlikely.
5. *Inconsistency*—T entails the negation of (a part of) T_1.

In principle, any relation short of full entailment (1) could be regarded as posing a conceptual problem for the theories exhibiting it. It should be stressed, however, that although situations (2) to (5) can generate conceptual problems, they *pose very different degress of cognitive threat;* those degrees are represented, in increasing order, by the sequence (2) through (5).

The Sources of Conceptual Problems

In discussing external conceptual problems, I was deliberately vague about what sorts of theories or beliefs can generate conceptual problems for a scientific theory. I have avoided this issue thus far because I wanted to focus first on the kinds of connections between theories which could generate conceptual problems. The time has come, however, to spell out the other side of the issue by asking what sorts of theories can qualify to be paired with a scientific theory in order to generate a conceptual problem; for unless we can answer that question coherently, one could trivially and mechanically generate conceptual problems for any theory simply by conjoining it arbitrarily with any "wild" belief we liked. For instance, we could create a problem for modern quantum theory by pointing out its lack of relevancy for Zen Buddhism! So far as I can tell, there are at least three distinct classes of difficulties which can generate external conceptual problems: (1) cases where two *scientific* theories from different domains are in tension; (2) cases where a scientific theory is in conflict with the *methodological* theories of the relevant scientific community; and (3) cases where a scientific theory is in conflict with any component of the prevalent *world view.* Each merits serious discussion.

Intra-scientific difficulties. It is very often the case that a new theory in some scientific domain will make assumptions about the world which are incompatible with the assumptions of another *scientific* theory, a theory which we have good independent grounds for accepting. Thus, the astronomical system of Copernicus—while not a theory of physics in itself—nonetheless made a number of assumptions about the motion of bodies which were inconsistent with the then accepted Aristotelian mechanics. One of the strongest sixteenth-century arguments against the Copernican system consisted in pointing out that the theory of Copernicus, although perhaps adequate so far as the astronomical evidence went, was unacceptable because it ran counter to the tenets of the best established physical theory. Even worse, Copernicus really had no well-articulated alternative system of mechanics with which to rationalize the assumptions he was making about the motion of the earth. It was Galileo's signal contribution to deal with this conceptual problem, by recognizing the incompatibility between Aristotelian physics and Copernican astronomy and by remedying the situation by designing a new physics which was independently plausible and compatible with Copernican astronomy.

The recognition and resolution of such conceptual problems has been one of the more fertile processes in the history of the natural and the social sciences.[10] If two scientific theories are inconsistent or mutually implausible, there is a strong presumption that at least one of them should be abandoned. That much is straightforward. What is more interesting is the fact that one generally cannot simply jettison one or the other of an inconsistent pair without wreaking havoc with the rest of scientific knowledge. Because theories in certain domains (say, astronomy) seem to require for their comprehension and empirical assessment the existence of theories in other domains (say, mechanics or optics),[11] *the decision to abandon one of a pair of inconsistent theories and to retain the other member of the pair usually involves a commitment to develop an adequate alternative to the rejected theory.*

As a result, such conceptual problems are generally much easier to recognize than to resolve. Rarely, if ever, can we resolve such problems by the simple device of rejecting one of the offending pair. Moreover, as we have already seen, there is nothing built into the process of scientific evaluation which can inform us in advance which member of an inconsistent pair ought to be rejected. That is a question which can be resolved only after the fact, i.e., once we have tried giving up one, then the other, and have observed with what success we can construct an adequate pair-member for the retained theory.

Two final points about intra-scientific conceptual problems should be made in passing. It should be stressed, first, that the fact that a particular theory is incompatible with another accepted theory creates a conceptual problem for *both* theories. The inconsistency relation is symmetrical, and we must not lose sight of the fact that intra-scientific conceptual problems inevitably raise presumptive doubts about both members of the incompatible pair. Second, we should observe that the noting of a logical inconsistency or a relation of non-reinforcement between two theories need *not* force scientists to abandon one, or the other, or both. Just as it can sometimes be rational to retain a theory in the face of anomalous evidence, so, too, can it be sometimes rational to retain a theory in the face of an inconsistency between it and some other accepted theory. What we must recognize is that the occurrence of such an inconsistency indicates a *weakness,* a reason for considering the abandonment of one or the other theory (or perhaps both).

Among the most vivid examples of intra-scientific difficulties were the controversies between biologists, geologists, and physicists in the late nineteenth century over the chronology of the earth. On the geological and biological side was an enormous amount of evidence to support the view that the earth was very old indeed, that it was partially fluid under the surface, and that physical conditions on its surface had remained largely unchanged for hundreds of millions of years. Both uniformitarian geology and evolutionary biology rested upon such assumptions. The physicist Lord Kelvin, however, found himself unable to reconcile these core postulates with thermodynamics. Specifically, he showed that the second law of thermodynamics (entailing an increase in entropy) was incompatible with an evolutionary account of species and that both the first and second laws were incompatible with the geologist's hypothesis that the energy reserves in the earth had remained constant through much of the geological past. General perplexity abounded. Thermodynamics had much going for it in physics, but the dominant geological and biological theories also could point to a huge reserve of solved problems. The dilemma was acute: ought one abandon thermodynamics, reject uniformitarian geology, or repudiate evolutionary theory? Or was there some other option? As it turned out, though no one could have foreseen this in advance, all three could be retained, since the discovery of radioactivity made it possible to circumvent the problems about energy conservation. What matters here, for our purposes, is that the emergence of this incompatibility created acute conceptual problems for *all* the sciences concerned. If the route to a resolution of the problems was murky, it was generally perceived that these conceptual problems, until resolved, raised strong doubts about the problem-solving efficacy of a wide range of scientific theories.

Normative difficulties. Science, as it is often said, is an activity, an activity conducted by seemingly rational agents. As such, it has certain aims and goals. The rational assessment of science must therefore be, in large measure, a matter of determining whether the theories of science achieve the cognitive goals of scientific activity. What are these goals and how do we achieve them? It is one of the central functions of any philosophy or methodology of science to specify those goals and to indicate the most effective *means* for achieving them. The whole point of a methodological rule (such as Newton's classic dictum, "hypotheses non fingo") is to offer a norm for scientific behavior; to tell us what we should, or should not, do in order to achieve the cognitive, epistemic, and practical goals of the scientific enterprise.

Since antiquity, philosophers and philosopher-scientists have sought to define sets of norms, or methodological rules, which are expected to govern the behavior of the scientist. From Aristotle to Ernst Mach, from Hippocrates to Claude Bernard, thinkers concerned about science have attempted to legislate concerning the acceptable modes of scientific inference. In the early seventeenth century, the dominant image was mathematical and demonstrative, an image that became canonical in Descartes' famous *Discourse on Method.* In the eighteenth and early nineteenth century, by contrast, most natural philosophers were convinced that the methods of science should be inductive and experimental. Not surprisingly, every historical epoch exhibits one or more dominant, normative images of science. It would be a serious mistake to imagine, as many historians do, that these norms are just the concern of the professional philosopher or logician. *Every* practicing scientist, past and present, adheres to certain views about how science should be performed, about what counts as an adequate explanation, about the use of experimental controls, and the like. *These norms, which a scientist brings to bear in his assessment of theories, have been perhaps the single major source for most of the controversies in the history of science, and for the generation of many of the most acute conceptual problems with which scientists have had to cope.*

It is still widely maintained that the methodology to which a scientist subscribes is really little more than perfunctory window dressing, which is honored more in the breach than in the observance. Prominent scientists and historical scholars of our own era (most notably Einstein and Koyré[12]) have scoffed at the idea that a scientist's explicit views about methodology can exert much impact on his scientific beliefs and activities. Moreover, there are significant cases (e.g., Newton and Galileo) in which a scientist's actual research violates almost every methodological rule to which he pays lip service. How, under those circumstances, can I argue here that methodology is a potent source for the evaluation of scientific theories and for the generation of conceptual problems?

Fortunately, the work of several historians in the last twenty years has provided overwhelming evidence that the methodological beliefs of scientists often do profoundly effect their research and their appraisals of the merits of scientific theories.[13] What all these investigations make clear (contra-Einstein and Koyré) is that the fate of most of the important scientific theories in the past have been closely bound up with the *methodological* appraisals of these theories; *methodological well-foundedness* has been constitutive of, rather than tangential to, the most important appraisals of theories.

It is for precisely that reason that perceived methodological weaknesses have constituted serious, and often acute, conceptual problems for any theory exhibiting them. It is for the same reason that the elimination of incompatibilities between a theory and the relevant methodology constitutes one of the most impressive ways in which a theory can improve its cognitive standing.

The resolution of a "tension" between a methodology and a scientific theory is often achieved by modifying the scientific theory so as to reconcile it to the methodological norms. But such problems are not always resolved in this fashion. In many cases, *it is the methodology itself which is altered.* Consider, as but one example, the development of Newtonian theory in the eighteenth century. By the 1720s, the dominant methodology accepted alike by scientists and philosophers was an *inductivist* one. Following the claims of Bacon, Locke, and Newton himself, researchers were convinced that the only legitimate theories were those which could be inductively inferred by simple generalization from observable data. Unfortunately, however, the direction of physical theory by the 1740s and 1750s scarcely seemed to square with this explicit inductivist methodology. Within electricity, heat theory, pneumatics, chemistry and physiology, Newtonian theories were emerging which postulated the existence of imperceptible particles and fluids—entities which could not conceivably be "inductively inferred" from observed data. The incompatibility of these new theories with the explicit methodology of the Newtonian research tradition produced acute conceptual problems. Some Newtonians (especially those in the so-called "Scottish School") sought to resolve the conceptual problems by simply repudiating those physical theories which violated the accepted methodological norms.[14] Other Newtonians (e.g., LeSage, Hartley, and Lambert) insisted *the norms themselves should be changed* so as to bring them into line with the best available physical theories.[15] This latter group took it on themselves to hammer out a new methodology for science which would provide a license for theorizing about unseen entities. (In its essentials, the methodology they produced was the hypothetico-deductive methodology, which even now remains the dominant one.) This new methodology, by providing a rationale for "micro-theorizing," eliminated what had been a major conceptual stumbling block to the acceptance of a wide range of Newtonian theories in the mid and late eighteenth century. (Here, as above, historians with purely empiricist models of science have completely missed the occurrence, let alone the significance, of these developments in the evolution of the Newtonian research tradition.)

Other cases of methodologically induced conceptual problems abound. Much of the debate about uniformitarian geology, much of the controversy about atomism, the bulk of the opposition to psychoanalysis and behaviorism, and many of the quarrels in quantum mechanics focus upon the methodological strengths and weaknesses of the scientific theories in question. Cases of this kind make it clear that the recognition of normative conceptual problems is a much more potent force in the historical evolution of science than some historians of science have recognized.

But if historians have sometimes underestimated the importance of such conceptual problems, their culpability is insignificant when compared to the utter failure of philosophers to find any role for this sort of problem in their accounts of scientific change. Even those philosophers who have been liberal enough to find a role for metaphysics in scientific development have completely ignored the fact that the methodology to which a scientist subscribes does, and should have, a major role to play in determining that scientist's assessment of the rational merits of competing scientific theories. If a scientist has good grounds for accepting some methodology and if some scientific theory violates that methodology, then it is entirely rational for him to have grave reservations about the theory. (It is one of the crueler ironies of recent epistemology that epistemologists themselves have never come to terms with, nor found a rationale for, the decisive role which epistemology and methodology have enjoyed in the rational development of the sciences.)

Worldview difficulties. The third type of external conceptual problem arises when a particular scientific theory is seen to be incompatible with, or not mutually reinforcing for, some other body of accepted, but *prima facie* nonscientific, beliefs. Within any culture, there are widely accepted beliefs which go beyond the scientific domain. Although the exact proportion of scientific and nonscientific propositions within the total population of reasonable beliefs changes with time, there has never been a period in the history of thought when the theories of science exhausted the domain of rational belief. What I am calling worldview difficulties are like intra-scientific difficulties, except that here the inconsistency, or lack of mutual reinforcement, is not *within* the framework of science itself, but rather between science and our "extra-scientific beliefs." Such beliefs fall in areas as diverse as metaphysics, logic, ethics and theology.

For example, one of the central conceptual problems confronting the Newtonians in the eighteenth century concerned the ontology of forces. How, critics such as Leibniz and Huygens had asked, can bodies exert force at points far removed from the bodies themselves? What substance carries the attractive force of the sun through 90 million miles of empty space so that the earth is pulled towards it? How, at the more prosaic level, can a magnet draw towards itself a piece of iron several inches away? Such phenomena seemed to defy the very logic of speaking about substances and properties since properties (e.g., the power of attraction) seemed to be capable of disentangling themselves from the material bodies of which they were the properties. As Buchdahl,[16] Heimann and McGuire[17] have convincingly argued, sorting out this issue became one of the central philosophical and scientific problems of the Enlightenment. Not satisfied with the Cotesian denial that this was an acute conceptual problem (Cotes was prepared to say that nature was generally unintelligible and that the unintelligibility of distance forces was no particular source of cognitive concern[18]), philosophers and scientists all over Europe began to re-evaluate such traditional issues as the nature of substance, the relations of properties to substances, and, particularly, the nature of our knowledge of substance. What resulted from this reappraisal at the hands of Kant, Priestley, Hutton, and others was a new ontology which argued for the priority of force over matter and which made the powers of activity (rather than passive powers like mass and inertia) into the basic building blocks of the physical world. The emergence of this new ontology did several things at once: it eliminated the most acute conceptual problem for Newtonian science by exhibiting the "intelligibility" of action-at-a-distance; it brought the ontology of philosophy and the ontology of physics back into harmony; and it made possible the subsequent emergence of theories of the physical field.[19]

Those "positivist" philosophers and historians of science who see the progress of science entirely in empirical terms have completely missed the huge significance of these developments for science as well as for philosophy. Convinced that metaphysics is foreign, even alien, to the development of scientific ideas, they have written about the history of Newtonianism without even perceiving the vital bearing of these metaphysical controversies on the historical career of Newtonian doctrines.

Traditionally, worldview difficulties have tended to emerge most often as a result of tensions between science, on the one hand, and either theology, philosophy or social theory, on the other hand.[20] It is well known, for instance, that one of the major difficulties for the mechanistic scientific program of the seventeenth and eighteenth centuries was the perceived discrepancy between a theory which reduced the cosmos to a self-operating machine and certain "activist" theologies which sought to preserve an important role to God in the day-to-day maintenance of the universe. The famous *Leibniz-Clarke Correspondence,* one of the major documents of the early Enlightenment, is replete with controversies that illustrate what I call world-view difficulties. Similarly, one major stumbling block to the emergence of evolutionary theory was the conviction, based on the best available philosophical insights, that species must be separate and distinct.[21] More recently, one of the most persistent set of conceptual problems in twentieth century physics has been the dissonance between quantum mechanics and our "philosophical" beliefs about causality, change, substance and "reality."

It is not only incompatibilities between science and philosophy or between science and theology which can lead to world-view difficulties. Conflicts with a social or moral ideology can produce similar tensions. In our own time, for instance, there are several instances where seemingly serious arguments have been lodged against a scientific theory because of moral or ethical worldview difficulties. In the Soviet Union, the Lysenko affair is a case in point. Because evolutionary biology, with its denial of the transmission of acquired characteristics, ran counter to the Marxist view that man's very nature could be changed by his environment, there were strong reservations voiced against Darwinism and Mendelism and much support was given to a scientific research effort like Lysenko's which sought to find scientific evidence for the Marxist philosophy of man. In the West, similar constraints have recently confronted researchers and theorists examining the possibility of racial differences. It has been suggested that any scientific theory which would argue for differences of ability or intelligence between the various races must necessarily be unsound, because such a doctrine runs counter to our egalitarian social and political framework.

There is a prominent group of thinkers in contemporary science and philosophy who have argued that world view difficulties are only pseudo-problems.[22] They claim that scientific theories can stand alone and that any element of our worldview which does not square with science should simply be abandoned. I shall take issue in the next chapter with this positivistic doctrine, but for now, I should make a few disclaimers, lest I be taken for asserting more than I am:

1. It is *not* my claim that a scientific theory should necessarily be abandoned when it encounters worldview problems; in asserting the existence of conceptual problems of this type, I am only asserting the *fact* that a tension often exists between our "scientific" beliefs and our "nonscientific" ones, and that such a tension does pose a problem for *both* sets of beliefs. How that tension is to be resolved depends on the particularities of the case.

2. It is *not* my claim that every worldview problem constitutes a serious ground for reservations about a scientific theory. How serious the problem is for the theory depends upon how well entrenched the nonscientific belief is and upon what problem-solving capabilities we would lose by abandoning it.

The Relative Weighting of Conceptual Problems

Having examined in a little more detail how conceptual problems are generated, we can now think about how to assess their relative importance. It is vital to stress, at the outset, that a conceptual problem will, in general, be a *more* serious one than an empirical anomaly. No one, for instance, proposed abandoning Newtonian mechanics when it could not accurately predict the motion of the moon. But many thinkers (such as Leibniz, Huygens, and Wolff) were seriously prepared to dismiss Newtonian physics because its ontology was incompatible with the accepted metaphysics of the day. This difference in weighting arises, not because science is more rationalistic than empirical; but rather because it is usually easier to explain away an anomalous experimental result than to dismiss out of hand a conceptual problem.[23] (Let me add that I am *not* suggesting that all conceptual problems are more important than all empirical problems. I am rather making the more modest claim that most conceptual problems are of greater moment than most empirical anomalies.)

Within the domain of conceptual problems, there are certain circumstances which tend to promote or demote the initial importance of such problems. There are at least four situations which should be distinguished here:

1. As we have already seen, the nature of the logical relation between two theories exhibiting a conceptual problem can vary enormously from inconsistency (in its most acute form) to mutual support. Other things being equal, the greater the tension between two theories, the weightier the problem will be.

2. When a conceptual problem arises as a result of a conflict between two theories, T_1 and T_2, the seriousness of that problem for T_1 depends on how confident we are about the acceptability of T_2. If T_2 has proven to be extremely effective at solving empirical problems and if its abandonment would leave us with many anomalies, then matters are very difficult for the proponents of T_1. If, on the other hand, T_2's record as a problem solver is very modest, then T_2's incompatibility with T_1 will probably not count as a major conceptual problem for T_1.

3. Another case in which it becomes meaningful to speak of the grading of conceptual problems on a scale of importance occurs when—within a particular scientific domain—we have two *competing* (as opposed to complementary) theories, T_1 and T_2. If both T_1 and T_2 exhibit the *same* conceptual problem(s), then those problems count no more against one than against the other and become relatively insignificant in the context of comparative theory appraisal. However, if T_1 generates certain conceptual problems which T_2 does *not,* then those problems become highly significant in the appraisal of the relative merits of T_1 and T_2.

4. A final determinant of the importance of a conceptual problem (as with anomalies) has to do with the "age" of that problem. If it has only recently been discovered that a theory poses a certain conceptual problem (for instance, an internal inconsistency), there is usually some grounds for hope that, with very minor modifications in the theory, we can bring it into line and thus eliminate the problem. The threat which the problem poses to the theory is generally offset by an optimism that it can be readily dealt with—an optimism that is often justified. If, on the other hand, a theory has been known to have a particular conceptual problem for some length of time, if partisans of that

theory have tried, repeatedly and unsuccessfully, to make the theory consistent, or to reconcile it with our norms and our other accepted beliefs, then that problem assumes an ever greater importance with time, and assumes an ever greater significance in debates about the acceptability of the theory (or theories) which generate(s) it.

Summary and Overview

Quite simply, the claim of this chapter is that *no* major contemporary philosophy of science allows scope for the weighty role which conceptual problems have played in the history of science. Even those philosophers who claim to take the actual evolution of science seriously (e.g., Lakatos, Kuhn, Feyerabend, and Hanson) have made no serious concessions to the non-empirical dimensions of scientific debate. We now know enough about the importance of these nonempirical factors within the evolution of science to say with some confidence that *any theory about the nature of science which finds no role for conceptual problems forfeits any claim to being a theory about how science has actually evolved.*

Although the analytic machinery thus far developed is still insufficent for constructing a general model of scientific progress and growth, we now possess enough pieces of the puzzle that we can begin to talk in an approximative way about what a problem-solving model of progress might look like. The core assumptions of such a model are simple ones: (1) *the solved problem*—empirical or conceptual—*is the basic unit of scientific progress;* and (2) *the aim of science is to maximize the scope of solved empirical problems, while minimizing the scope of anomalous and conceptual problems.*

The more numerous and weightier the problems are which a theory can adequately solve, the better it is. If one theory can solve more significant problems than a competitor, then it is preferable to it. At one level, this is a noncontroversial claim. If we interpret problems exclusively in the sense of what we have called "solved empirical problems," many philosophers of science would accept that progress does amount to the solution of such problems. But, as we have seen, *there are problems in science other than solved empirical ones, specifically anomalous and conceptual problems.* My definition of progress chiefly becomes controversial (and potentially interesting) when we interpret it as applying to the latter as well as to the former. My reasons for wishing to broaden the base in this way should now be clear. If it counts in favor of a theory when it can accumulate solved empirical problems (as the standard view allows), then it should also count *against* a theory if it generates anomalous and conceptual problems. *Indeed, the problem-solving effectiveness of a theory depends on the balance it strikes between its solved problems and its unresolved problems.* How exactly does this work?

Let us begin with a very crude model of scientific evolution. Imagine some domain in which we notice a certain puzzling phenomenon, *p*. The phenomenon *p* constitutes an unsolved problem for the scientist who wishes to develop a theory, T_1, specifically with a view toward resolving *p*. Once T_1 is announced, several things are likely to happen simultaneously. Some fellow scientist may observe that T_1 predicts other phenomena in the domain besides *p*. These predictions will be tested, and, very often, some of them will not be borne out in our observation. Thus, the observation of these discrepant results will constitute one or more anomalies for T_1. At the same time, it may be pointed out that T_1 makes certain assumptions about natural processes which run counter to some of our most widely accepted theories, or that it is incompatible with our methodological norms. This will constitute one or more conceptual problems for theory T_1.

Thus far in this imaginary chronology, we are not clear whether any progress has been made. It is true that T_1 has solved its initial empirical problem, *p*, and to that extent, we can say that "progress" has been made. Unfortunately, however, the very theory, T_1, which cleared up that problem, has generated several others; in this case, anomalies and conceptual problems. It is entirely possible that more serious problems have been generated than resolved by the invention of T_1. But let us carry the example through in time for a while. Suppose that a second theorist comes along who is convinced that he can improve T_1. What does improving T_1 mean? Very roughly, such improvement would be exhibited by showing that a new theory, T_2, could explain the initial empirical problem of T_1 without generating the same, or as many, anomalies and conceptual problems as T_1 produced. If T_2 managed to do as much work at the empirical problem level as T_1 did, without all of T_1's attendant empirical and conceptual difficulties, we could all agree that it would be more reasonable to accept T_2 than to accept T_1; that, indeed, the acceptance of T_2 was progressive and that the continued espousal of T_1 was unprogressive or regressive.

Generalizing from this simple example, we could define an *appraisal measure* for a theory in the following way: *the overall problem-solving effectiveness of a theory is determined by assessing the number and importance of the empirical problems which the theory solves and deducting therefrom the number and importance of the anomalies and conceptual problems which the theory generates.*

The step from here to a rudimentary notion of scientific progress is straightforward. Given that the aim of science is problem solving (or, more precisely, the mini-max strategy sketched above), *progress can occur if and only if the succession of scientific theories in any domain shows an increasing degree of problem solving effectiveness.* Localizing the notion of progress to specific situations rather than to large stretches of time, we can say that *any time we modify a theory or replace it by another theory, that change is progressive if and only if the later version is a more effective problem solver (in the sense just defined) than its predecessor.*

There are many ways in which such progress can occur. It may come about simply by an expansion of the domain of solved empirical problems with all the other appraisal vectors remaining fixed. In such a case, the replacement of T_1 by T_2 (which solves more empirical problems) is clearly progressive. Progress can also result from a modification of the theory which eliminates some troublesome anomalies or which resolves some conceptual problems. Most often, of course, progress occurs as a result of all the relevant variables shifting subtly.

Given the exclusive emphasis by most philosophers on empirical problems, and their solution, it is important to stress that, on the model outlined here, (1) progress can occur without an expansion of the domain of solved empirical problems, and is even conceivable when the domain of such problems contracts; and (2) a theory change may conceivably be non-progressive or *regressive,* even when the index of solved empirical problems *increases,* specifically, if the change leads to more acute anomalies or conceptual problems confronting the new theory than those exhibited by the predecessor theory.

Although an outline of a theory of cognitive progress is now emerging, there is still one crucial dimension missing. In all the talk about problem solving, there has been some confusion about what *kinds* of things solve problems. I have been using the term "theory" to designate those complexes whose problem-solving capacities must be appraised; in order to clarify the types of problems in science, I have had to postpone a discussion about what kinds of things can solve problems. We must examine that side of the problem-solving equation before the rough-hewn model of progress outlined here can be refined into a valuable tool of analysis.

Chapter Three

From Theories to Research Traditions

The intellectual function of an established conceptual scheme is to determine the patterns of theory, the meaningful questions, the legitimate interpretations . . . S. TOULMIN (1970), p. 40

Theories are inevitably involved in the solution of problems; the very aim of theorizing is to provide coherent and adequate solutions to the empirical problems which stimulate inquiry. Theories, moreover, are designed to avoid (or to resolve) the various conceptual and anomalous problems which their predecessors generate. If one looks at inquiry in this way, if one views theories from this perspective, it becomes clear that *the central cognitive test of any theory involves assessing its adequacy as a solution of certain empirical and conceptual problems.* Having developed in earlier chapters a taxonomy for describing the kinds of problems which confront theories, we must now lay down adequacy conditions for determining when a theory provides an acceptable solution to the problems which confront it.

But before we embark on that task, we must clarify what theories are and how they function, for a failure to make some rudimentary distinctions here has brought grief to more than one major philosophy of science. Entire books have been devoted to the structure of scientific theory; I am attempting nothing that ambitious. Rather, I shall want to insist on only two major points with respect to an analysis of theories.

In the first place, to make explicit what has been implicit all along, *the evaluation of theories is a comparative matter.* What is crucial in any cognitive assessment of a theory is how it fares with respect to its competitors. Absolute measures of the empirical or conceptual credentials of a theory are of no significance; decisive is the judgment as to how a theory stacks up against its known contenders. Much of the literature in the philosophy of science has been based upon the assumption that theoretical evaluation occurs in a competitive vacuum. By contrast, I shall be assuming that assessments of theories always involve comparative modalities. We ask: is this theory better than that one? Is this doctrine the best among the available options?

The second major claim of this chapter is that *it is necessary to distinguish, within the class of what are usually called "scientific theories," between two different sorts of propositional networks.*

In the standard literature on scientific inference, as well as in common scientific practice, the term "theory" refers to (at least) two very types of things. We often use the term "theory" to denote a very specific set of related doctrines (commonly called "hypotheses" or "axioms" or "principles") which can be utilized for making specific experimental predictions and for giving detailed explanations of natural phenomena. Examples of this type of theory would include Maxwell's theory of electromagnetism, the Bohr-Kramers-Slater theory of atomic structure, Einstein's theory of the photoelectric effect, Marx's labor theory of value, Wegener's theory of continental drift, and the Freudian theory of the Oedipal complex.

By contrast, the term "theory" is also used to refer to much more general, much less easily testable, sets of doctrines or assumptions. For instance, one speaks about "the atomic theory," or "the theory of evolution," or "the kinetic theory of gases." In each of these cases, we are referring not to a single theory, but to a whole spectrum of individual theories. The term "evolutionary theory" for instance, does not refer to any single theory but to an entire family of doctrines, historically and conceptually related, all of which work from the assumption that organic species have common lines of descent. Similarly, the term "atomic theory" generally refers to a large set of doctrines, all of which are predicated on the assumption that matter is discontinuous. A particularly vivid instance of one theory which includes a wide variety of specific instantiations is offered by recent "quantum theory." Since 1930, that term has included (among other things) quantum field theories, group theories, so-called S-matrix theories, and renormalized field theories—between any two of which there are huge conceptual divergences.

The differences between the two types of theories outlined above are vast: not only are there contrasts of generality and specificity between them, but the modes of appraisal and evaluation appropriate to each are radically different. It will be the central claim of this chapter that *until we become mindful of the cognitive and evaluational differences between these two types of theories, it will be impossible to have a theory of scientific progress which is historically sound or philosophically adequate.*

But it is not only fidelity to scientific practice and usage which requires us to take these larger theoretical units seriously. Much of the research done by historians and philosophers of science in the last decade suggests that these more general units of analysis exhibit many of the epistemic features which, although most characteristic of science, elude the analyst who limits his range to theories in the narrower sense. Specifically, it has been suggested by Kuhn and Lakatos that *the more general theories,* rather than the more specific ones, *are the primary tool for understanding and appraising scientific progress.*

I share this conviction in principle, but find that the accounts hitherto given of what these larger theories are, and how they evolve, are not fully satisfactory. Because the bulk of this chapter will be devoted to outlining a new account of the more global theories (which I shall be calling *research traditions),* it is appropriate that I should indicate what I find chiefly wanting in the best known efforts to grapple with this problem. Of the many theories of scientific evolution that have been developed, two specifically address themselves to the question of the nature of these more general theories.

Kuhn's Theory of Scientific "Paradigms"

In his influential *Structure of Scientific Revolutions,* Thomas Kuhn offers a model of scientific progress whose primary element is the "paradigm." Although Kuhn's notion of paradigms has been shown to be systematically ambiguous[1] (and thus difficult to characterize accurately), they do have certain identifiable characteristics. They are, to begin with, "ways of looking at the world"; broad quasi-metaphysical insights or hunches about how the phenomena in some domain should be explained. Included under the umbrella of any well-developed paradigm will be a number of specific theories, each of which presupposes one or more elements of the paradigm. Once a paradigm is accepted by scientists (and one of Kuhn's more extreme claims is that in any "mature" science,[2] *every* scientist will accept the *same* paradigm most of the time), they can proceed with the process of "paradigm articulation," also known as "normal science." In periods of normal science, the dominant paradigm will itself be regarded as unalterable and immune from criticism. Individual, specific theories (which represent efforts "to articulate the paradigm," i.e., to apply it to an ever wider range of cases) may well be criticized, falsified and abandoned; but the paradigm itself is unchallenged. It remains so until enough "anomalies"[3] accumulate (Kuhn never indicates how this point is determined) that scientists begin to ask whether the dominant paradigm is really appropriate. Kuhn calls this time a period of "crisis." During a crisis, scientists begin for the first time to consider seriously alternative paradigms. If one of those alternatives proves to be more *empirically successful* than the former paradigm, a scientific revolution occurs, a new paradigm is enthroned, and another period of normal science ensues.

There is much that is valuable in Kuhn's approach. He recognizes clearly that maxi-theories have different cognitive and heuristic functions than mini-theories. He has probably been the first thinker to stress the tenacity and persevering qualities of global theories—even when confronted with serious anomalies.[4] He has correctly rejected the (widely assumed) cumulative character of science.[5] But for all its many strengths, Kuhn's model of scientific progress suffers from some acute conceptual and empirical difficulties. For instance, Kuhn's account of paradigms and their careers has been extensively criticized by Shapere, who has highlighted the obscure and opaque character of the paradigm itself by pointing out many inconsistencies in Kuhn's use of the notion.[6] Feyerabend[7] and others have stressed the historical incorrectness of Kuhn's stipulation that "normal science" is in any way typical or normal. Virtually every major period in the history of science is characterized both by the co-existence of numerous competing paradigms, with none exerting hegemony over the field, and by the persistent and continuous manner in which the foundational assumptions of every paradigm are debated within the scientific community. Numerous critics have noted the arbitrariness of Kuhn's theory of crisis: if (as Kuhn says) a few anomalies do not produce a crisis, but "many" do, how does the scientist determine the "crisis point?" There are other serious flaws as well. In my view, the most significant of these are:

1. Kuhn's failure to see *the role of conceptual problems* in scientific debate and in paradigm evaluation. Insofar as Kuhn grants that there are any rational criteria for paradigm choice, or for assessing the "progressiveness" of a paradigm, those criteria are the traditional positivist ones such as: Does the theory explain more facts than its predecessor? Can it solve some empirical anomalies exhibited by its predecessor? The whole notion of conceptual problems and their connection with progress finds no serious exemplification in Kuhn's analysis.

2. Kuhn never really resolves the crucial question of *the relationship between a paradigm and its constituent theories.* Does the paradigm entail or merely inspire its constituent theories? Do these theories, once developed, justify the paradigm, or does the paradigm justify them? It is not even clear,

in Kuhn's case, whether a paradigm precedes its theories or arises *nolens volens* after their formulation. Although this issue is extremely complex, any adequate theory of science is going to have to come to grips with it more directly than Kuhn has.

3. Kuhn's paradigms have a rigidity of structure which precludes them from evolving through the course of time in response to the weaknesses and anomalies which they generate. Moreover, because he makes the core assumptions of the paradigm immune from criticism, *there can be no corrective relationship between the paradigm and the data.* Accordingly, it is very difficult to square the inflexibility of Kuhnian paradigms with the historical fact that many maxi-theories have evolved through time.

4. Kuhn's paradigms, or "disciplinary matrices," are always implicit, never fully articulated.[8] As a result, it is difficult to understand how he can account for the many theoretical controversies which have occurred in the development of science, since scientists can presumably only debate about assumptions which have been made reasonably explicit. When, for instance, a Kuhnian maintains that the ontological and methodological frameworks for Cartesian or Newtonian physics, for Darwinian biology, or for behavioristic psychology were only implicit and never received overt formulation, he is running squarely in the face of the historical fact that the core assumptions of all these paradigms were explicit even from their inception.

5. Because paradigms are so implicit and can only be identified by pointing to their "exemplars" (basically an archetypal application of a mathematical formulation to an experimental problem), it follows that whenever two scientists utilize the same exemplars, they are, for Kuhn, *ipso facto* committed to the same paradigm. Such an approach ignores the persistent fact that different scientists often utilize the same laws or exemplars, yet subscribe to radically divergent views about the most basic questions of scientific ontology and methodology. (For instance, both mechanists and energeticists accepted identical conservation laws.) To this extent, analysing science in terms of paradigms is unlikely to reveal that "strong

network of commitments—conceptual, theoretical, instrumental, and metaphysical"[9] which Kuhn hoped to localize with his theory of paradigms.

Lakatos' Theory of "Research Programmes"

Largely in response to Kuhn's assault on some of the cherished assumptions of traditional philosophy of science, Imre Lakatos has developed an alternative theory about the role of these "super-theories" in the evolution of science. Calling such general theories "research programmes," Lakatos argues that research programmes have three elements: (1) a "hard-core" (or "negative heuristic") of fundamental assumptions which cannot be abandoned or modified without repudiation of the research programme;[10] (2) the "positive heuristic," which contains "a partially articulated set of suggestions or hints on how to change, . . . modify, sophisticate [*sic*]"[11] our specific theories whenever we wish to improve them, and (3) "a series of theories, T_1, T_2, T_3, . . ." where each subsequent theory "results from adding auxiliary clauses to . . . the previous theory."[12] Such theories are the specific instantiations of the general research programme. Research programmes can be progressive or regressive in a variety of ways: but progress, for Lakatos even more than for Kuhn, is a function exclusively of the *empirical* growth of a tradition. It is the possession of greater "empirical content," or of a higher "degree of empirical corroboration" which makes one theory superior to, and more progressive than, another.

Lakatos' model is, in many respects, a decided improvement on Kuhn's. Unlike Kuhn, Lakatos allows for, and stresses, the historical importance of the co-existence of several alternative research programmes at the same time, within the same domain. Unlike Kuhn, who often takes the view that paradigms are incommensurable[13] and thus not open to rational comparison, Lakatos insists that we can objectively compare the relative progress of competing research traditions. More than Kuhn, Lakatos tries to grapple with the thorny question of the relation of the super-theory to its constituent mini-theories.

But against that, Lakatos' model of research programmes shares many of the flaws of Kuhn's paradigms, and introduces some new ones as well:

1. As with Kuhn, Lakatos' conception of progress is exclusively empirical; the only progressive modifications in a theory are those which increase the scope of its empirical claims.

2. The sorts of changes which Lakatos allows within the mini-theories which constitute his research programme are extremely restricted. In essence, Lakatos only permits, as the relation between any theory and its successor within a research programme, the addition of a new assumption or a semantic re-interpretation of terms in the predecessor theory. On this remarkable view of things, *two theories can only be in the same research programme if one of the two entails the other.* As we shall see shortly, in the vast majority of cases, the succession of specific theories within a maxi-theory involves the *elimination* as well as the addition of assumptions, and there are rarely successor theories which entail their predecessors.

3. A fatal flaw in the Lakatosian notion of research programmes is its dependence upon the Tarski-Popper notions of "empirical and logical content." *All* Lakatos' measures of progress require a comparison of the empirical content of every member of the series of theories which constitutes any research programme.[14] As Grünbaum and others have shown convincingly, the attempt to specify content measures for scientific theories is extremely problematic if not literally impossible.[15] Because comparisons of content are generally impossible, neither Lakatos nor his followers have been able to identify *any* historical case to which the Lakatosian definition of progress can be shown strictly to apply.[16]

4. Because of Lakatos' idiosyncratic view that the acceptance of theories can scarcely if ever be rational, he cannot translate his assessments of progress (assuming he could make them!) into recommendations about cognitive action.[17] Although one research programme may be more progressive than another, we can, on Lakatos' account, deduce nothing from that about which research programme should be preferred or accepted. As a result, there can never be a connection between a theory

of progress and a theory of rational acceptability (or, to use Lakatos' language, between methodological "appraisal" and "advice").

5. Lakatos' claim that the accumulation of anomalies has no bearing on the appraisal of a research programme is massively refuted by the history of science.

6. Lakatos' research programmes, like Kuhn's paradigms, are rigid in their hard-core structure and admit of no fundamental changes.[18]

What should be clear, even from this very brief survey of two of the major theories of scientific change, is that there are a number of analytical and historical difficulties confronting existing attempts to understand the nature and role of maxi-theories. With some of those difficulties in mind, we can turn now to explore an alternative model of scientific progress, built upon elements outlined in the previous chapters. One crucial test of that model will be whether it can avoid some of the problems which handicap its predecessors. Although there are numerous common elements between my model and those of Kuhn and Lakatos (and I readily concede a great debt to their pioneering work), there are a sufficiently large number of differences that I shall try to develop the notion of a research tradition more or less from scratch.

The Nature of Research Traditions

We have already referred to a few classic research traditions: Darwinism, quantum theory, the electromagnetic theory of light. Every intellectual discipline, scientific as well as nonscientific, has a history replete with research traditions: empiricism and nominalism in philosophy, voluntarism and necessitarianism in theology, behaviorism and Freudianism in psychology, utilitarianism and intuitionism in ethics, Marxism and capitalism in economics, mechanism and vitalism in physiology, to name only a few. Such research traditions have a number of common traits:

1. Every research tradition has a number of specific theories which exemplify and partially constitute it; some of these theories will be contemporaneous, others will be temporal successors of earlier ones;

2. Every research tradition exhibits certain *metaphysical* and *methodological* commitments which, as an ensemble, individuate the research tradition and distinguish it from others;

3. Each research tradition (unlike a specific theory) goes through a number of different, detailed (and often mutually contradictory) formulations and generally has a long history extending through a significant period of time. (By contrast, theories are frequently short-lived.)

These are by no means the only important characteristics of research traditions, but they should serve, for the time being, to identify the kinds of objects whose properties I would like to explore.

In brief, a research tradition provides a set of guidelines for the development of specific theories. Part of those guidelines constitute an ontology which specifies, in a general way, the types of fundamental entities which exist in the domain or domains within which the research tradition is embedded. The function of specific theories within the research tradition is to explain all the empirical problems in the domain by "reducing" them to the ontology of the research tradition. If the research tradition is behaviorism, for instance, it tells us that the only legitimate entities which behavioristic theories can postulate are directly and publicly observable physical and physiological signs. If the research tradition is that of Cartesian physics, it specifies that only matter and minds exist, and that theories which talk of other types of substances (or of "mixed" mind and matter) are unacceptable. Moreover, the research tradition *outlines the different modes by which these entities can interact.* Thus, Cartesian particles can only interact by contact, not by action-at-a-distance. Entities, within a Marxist research tradition, can only interact by virtue of the economic forces influencing them.

Very often, the research tradition will also specify certain modes of procedure which constitute the legitimate *methods of inquiry* open to a researcher within that tradition. These methodological principles will be wide-ranging in scope, addressing themselves to experimental techniques, modes of theoretical testing and evaluation, and the like. For instance, the methodological posture of the scientist in a strict Newtonian research tradition is inevitably inductivist, allowing for the espousal of only those theories which have been "inductively inferred" from the data. The methods of procedure outlined for a behavioristic psychologist are what is usually called "operationalist." Put simplistically, *a research tradition is thus a set of ontological and methodological "do's" and "don'ts."* To attempt what is forbidden by the metaphysics and methodology of a research tradition is to put oneself outside that tradition and to repudiate it. If, for instance, a Cartesian physicist starts talking about forces acting-at-a-distance, if a behaviorist starts talking about subconscious drives, if a Marxist begins speculating about ideas which do not arise in response to the economic substructure; in each of these cases, the activity indicated puts the scientist in question beyond the pale. By breaking with the ontology or the methodology of the research tradition within which he has worked, he has violated the strictures of that research tradition and divorced himself from it. Needless to say, that is not necessarily a bad thing. Some of the most important revolutions in scientific thought have come from thinkers who had the ingenuity to break with the research traditions of their day and to inaugurate new ones. But what we must preserve, if we are to understand either the logic or the history of the natural sciences, is the notion of the *integrity* of a reseach tradition, for it is precisely that integrity which stimulates, defines and delimits what can count as a solution to many of the most important scientific problems.[19]

Although it is vital to distinguish between the ontological and the methodological components of a research tradition, the two are often intimately related, and for a very natural reason: namely, that one's views about the appropriate *methods* of inquiry are generally compatible with one's views about the *objects* of inquiry. When, for instance, Charles Lyell defined the "uniformitarian" research tradition in geology, his ontology was restricted to presently acting causes and his methodology insisted that we should "explain past effects in terms of presently acting causes." Without a "presentist" ontology, his uniformitarian methodology would have been inappropriate; and without the latter, the presentist ontology would not have allowed Lyell to explain the geological past. Similarly, the mathematical ontology of the Cartesian research tradition (an ontology which argued that *all* physical changes were entirely changes of *quantity)* was very closely connected with the (mathematically inspired) deductivist and axiomatic methodology of Cartesianism. As we shall see later, it does not always happen that the ontology and methodology of a research tradition are so closely intertwined (for instance, the inductivist methodology of the Newtonian research tradition had only the weakest of connections with that tradition's ontology), but such cases are the exception rather than the rule.

So a preliminary, working definition of a research tradition could be put as follows: *a research tradition is a set of general assumptions about the entities and processes in a domain of study, and about the appropriate methods to be used for investigating the problems and constructing the theories in that domain.*

Theories and Research Traditions

Every research tradition will be associated with a series of specific theories, each of which is designed to particularize the ontology of the research tradition and to illustrate, or satisfy, its methodology. The mechanistic research tradition in seventeenth century optics, for example, includes several of Descartes' theories as well as the optical theories of Hooke, Rohault, Hobbes, Régis, and Huygens.[20] The phlogistic tradition in eighteenth century chemistry received more than a dozen specific theoretical formulations.[21] *Many of the theories within any evolving research tradition will be mutually inconsistent rivals,* precisely because some theories represent attempts, within the framework of the tradition, to improve and correct their predecessors.

The individual theories constituting the tradition will generally be empirically testable for they will entail (in conjunction with other specific theories) some precise predictions about how objects in the domain will behave. By contrast, *research traditions are neither explanatory, nor predictive, nor directly testable.* Their very generality, as well as their normative elements, precludes them from leading to detailed accounts of specific natural processes.

Except at the abstract level of specifying what the world is made of, and how it should be studied, research traditions do not provide detailed answers to specific questions. A research tradition will not tell us what happens to light when it is refracted at an interface between water and air; it will not tell us what happens if we put an eight-month-old female mouse into a maze; it will not tell us why lead melts at a lower temperature than copper. But it would be a mistake to conclude from the fact that research traditions do not offer solutions to specific problems that they are outside of the problem-solving process. To the contrary, the whole function of a research tradition is to provide us with the crucial tools we need for solving problems, both empirical and conceptual. (As we shall see later, the research tradition even goes so far as to define partially what the problems are, and what importance should be attached to them.) It is for just this reason that the objective evaluation of any research tradition is crucially linked with the problem-solving process. The very idea that an entity like a research tradition—which makes no predictions, which solves no specific problems, which is fundamentally normative and metaphysical—could be objectively evaluated may seem paradoxical. But nothing could be further from the case, for we can say quite simply that *a successful research tradition is one which leads, via its component theories, to the adequate solution of an increasing range of empirical and conceptual problems.* Determining whether a tradition is successful in this sense does not mean, of course, that the tradition has been "confirmed" or "refuted." Nor can such an appraisal tell us anything about the *truth* or *falsity* of the tradition.[22] A research tradition may be enormously successful at generating fruitful theories and yet flawed in its ontology or methodology. Equally, one can conceive that a research tradition might be true, and yet (perhaps because of the unimaginativeness of its proponents) unsuccessful at generating theories which were effective problem solvers. Hence to abandon or reject a reseach tradition is not (or ought not be) to judge that tradition false. Nor, in rejecting a research tradition as momentarily unsuccessful, are we necessarily relegating it to permanent oblivion; to the contrary, we can explicitly stipulate conditions which, if satisfied, would revive and recussitate it. Thus, when we reject a research tradition, we are merely making a tentative decision not to utilize it for the moment because there is an alternative to it which has proven to be a more successful problem solver.

Just as the fortunes of a research tradition are linked closely to the problem-solving effectiveness of its constituent theories, so too is the determination of the adequacy of a specific theory inextricably bound up with an assessment of the problem-solving effectiveness of the entire set of theories spawned by the research tradition of which that theory is a part.[23] If a theory is closely linked to an unsuccessful research tradition, then—whatever the problem-solving merits of that particular theory—it is likely to be regarded as highly suspect. For instance, Count Rumford's theories of heat conduction and convection were far superior to any alternative theories of thermal flow in fluids available in the period from 1800 to 1815. Nonetheless, few scientists took Rumford's theories seriously because (as they saw it) the research tradition in which Rumford worked (deriving from Boerhaave) had been discredited by the emergence of rival research traditions in chemistry (especially Joseph Black's), which suggested that heat was a substance rather than, as Rumford imagined, the random motion of particles. Rumford's specific theories only became fashionable in the 1840s and 1850s because by that time the balance between various research traditions had shifted sufficiently that many scientists were more prepared to consider seriously specific theories (like Rumford's) which grew out of a kinetic research tradition.

Contrariwise, a theory, even an inadequate one, will have some strong arguments in its favor if it is linked with a research tradition that is otherwise highly successful. Thus, theories of mechanistic physiology in the late seventeenth century (such as those of Borelli and Pitcairn) were highly regarded in many circles where the mechanistic research tradition was flourishing even though, judged entirely on their own merits, they were significantly inferior to certain theories in other, less successful research traditions.[24]

Up to this point, I have been deliberately vague about describing the kind of relation which exists between a theory and its "parent" research tradition. I have spoken of research traditions "inspiring" or "containing" or "generating" theories, and about theories "presupposing" or "constituting" or even "defining" research traditions. This is an extremely complex matter; the ambiguity of the metaphors I have invoked to characterize the theory/research tradition connection is a symptom of the difficulty of tackling this problem head-on.

But that task cannot be further postponed. I shall begin by saying what the relation between theories and research traditions is *not.* It is *not,* for instance, one of *entailment.* Research traditions do not entail their component theories; nor do those theories, taken either singly or jointly, entail their parent research traditions. One might wish it were otherwise, for then it would be a simple matter to determine mechanically which theories belonged to any given research tradition, or the research tradition(s) lurking behind any theory. But to see the theory/research tradition connection in such formal terms is completely to misunderstand the differences in kind between the two. A research tradition, at best, specifies a *general* ontology for nature, and a *general* method for solving natural problems within a given natural domain. A theory, on the other hand, articulates a very specific ontology and a number of specific and testable laws about nature. To be told, as the Newtonian research tradition in mechanics tells us, that we should treat all nonrectilinear motions as cases of centrally directed forces, does not entail *any* specific theory about how to explain, say, the motion of a compass needle in the vicinity of a current-carrying wire. To develop a "Newtonian" theory for that particular phenomenon, we must (as Ampère did) go far beyond the deductive consequences of the Newtonian research tradition. To be told, as the nineteenth-century "mechanical" research tradition tells us, that heat is simply a form of motion, does not deductively lead us to Boltzmann's version of the kinetic theory of gases or to statistical thermodynamics.

Similar considerations apply to the reverse relation between theories and research traditions. For example, given the theory of impact as developed by Huygens, we cannot deduce the basic assumptions of the research tradition within which Huygens worked. (We may, of course, be able to deduce that Huygens was working in a research tradition in which collision phenomena constituted an important unsolved problem, for if not, why should Huygens have bothered working out a theory of collision?). But it is never possible to deduce the whole of a research tradition from one, or even all, of the theories allied to it.

The reason why entailment will not help here is very simple: *there are a number of mutually inconsistent theories which can claim allegiance to the same research tradition, and there are a number of different research traditions which can, in principle, provide the presuppositional base for any given theory.*

Examples of both phenomena abound: many scientists in the Cartesian optical tradition argued that light travelled faster in optically denser media; other theorists, within the *same* tradition, asserted the converse. Staying within the history of optics, there are numerous examples of competing research traditions claiming to justify the same theory. For instance, Newton's theory that light has certain periodic properties was accepted alike by scientists in the wave and corpuscular traditions. If entailment were the relation between research traditions and theories, then it would be impossible for such situations to arise. Since the relation we are trying to explore is evidently not one of entailment, what can we say positively about it?

There are at least two specific modes by which theories and research traditions are related: one is *historical,* the other is *conceptual.* It is a matter of historical fact that most if not all of the major theories of science have emerged when the scientist who invented them was working within one or another specific research tradition. Boyle's theory of gases developed within the framework of the mechanical philosophy. Buffon's embryological theories were developed as efforts to apply the Newtonian research tradition to biological phenomena. Hartley's theories of sensation were developed within the research tradition of associationist psychology. Hertz's electrical theories were linked in important ways with the Maxwellian research tradition.

A specific theory, abstracted from its historical context, may not give unambiguous clues as to the research tradition (or traditions) with which it is associated. It is just this fact which has lead many scientists and philosophers to imagine that theories are usually appraised and assessed independently of the research traditions of which they are a part. But we should not be misled by the fact that a theory, taken abstractly, does not have its "parent" research tradition stamped all over it. Historical research can always (at least in principle) identify the research tradition(s) with which a particular theory has been associated. In this sense, the connection between a theory and a research tradition is as real as any fact of the past, and it is as important as the most important facts of the past. In order to see how important these connections are, we need to look at the ways in which theories and research traditions can interact.

The most important modes of interaction are generally influences of the research tradition upon its constituent theories. These influences take a variety of forms:

The problem determining role of research traditions. Even before specific theories are formulated within a tradition, and continuously thereafter, a research tradition will often strongly influence (though it does not *fully* determine) the range and the weighting of the empirical problems with which its component theories must grapple. Equally, research traditions have a decisive influence on what can count as the range of possible conceptual problems which the theories in that tradition can generate. These two processes are important ones and should be discussed in some detail.

1. Among the other roles of a research tradition, it is designed to delimit, at least partially and in outline, *the domain of application* of its constituent theories. It does this by indicating that it is appropriate to discuss certain classes of empirical problems in the given domain, whilst others belong to foreign domains, or are "pseudo-problems" which can be legitimately ignored. Either the ontology or the methodology of the research tradition can influence what are to count as legitimate problems for its constituent theories. If, for instance, the *methodology* of a research tradition specifies—as it usually

will—certain experimental techniques which alone are the legitimate investigational modes for determining what are the data to be explained, then it is clear that only "phenomena" which can be explored by those means can, in principle, count as legitimate empirical problems for theories within that tradition. A classic example of this process is offered by nineteenth-century phenomenological chemistry. Scientists in this tradition argued that the only legitimate problems to be solved by the chemist were those which concerned the *observable* reactions of chemcial reagents. Thus, to ask how this acid and this base react to form this salt is to pose an authentic problem. But to ask how atoms combine to form diatomic molecules cannot conceivably count as an empirical problem because the methodology of the research tradition denies the possibility of empirical knowledge of entities the size of atoms and molecules. For other research traditions in nineteenth-century chemistry, questions about the combining properties of certain entities not directly observable constituted authentic problems for empirical research.[25] (Contemporary behavioristic psychology and quantum mechanics likewise have methodologies which strongly preclude from consideration as problems certain "phenomena" which other research traditions countenance.)

Similarly, the *ontology* of a research tradition may exclude certain situations from, or include them within, the appropriate domain. Thus, the rise of the Cartesian mechanistic research tradition in the seventeenth century radically transformed the accepted problem domain for optical theories. It did so by arguing, or rather by simply postulating, that problems of perception and vision—problems which had classically been regarded as legitimate empirical problems for any optical theory—should be relegated to psychology and to physiology, fields outside the domain of optics, so that such empirical problems could be safely ignored by the mechanistic optical theorist.

A different kind of example is provided by late nineteenth-century physics, where the subtle fluid tradition (of Faraday, Maxwell, Hertz, and others) countenanced as legitimate empirical problems queries about the properties of the electromagnetic aether. Indeed, the classic Michelson-Morley experiments were originally conducted in order to determine the drag coefficient of bodies moving through such an aether. With the emergence of special relativity theory, however, a new research tradition and its related ontology cut out from the domain of the empirical problem of physics all questions about the elasticity, density, and velocity of the aether—questions which had been central *empirical* problems between 1850 and 1900.[26] These few examples should make it clear that research traditions can play a decisive role in specifying the sorts of things that are to count as potentially solvable empirical problems for their constituent theories.

2. Equally important is the way in which a research tradition can generate conceptual problems for its constituent theories. Indeed, the bulk of the conceptual problems which any theory may face will arise because of tensions between that theory and the research tradition of which it is a part. It often happens that the detailed articulation of a theory will lead to the adoption of assumptions which run counter to those allowed by the research tradtion of that theory. In such a situation, it is commonplace for critics of the theory to point to such a tension as a major conceptual problem for it. When, for instance, Huygens came to develop a general theory of motion, he found that the only empirically satisfactory theories were those which assumed vacua in nature. Unfortunately, Huygens was working squarely within the Cartesian research tradition, a tradition which identified space and matter and thus forbade empty spaces. As Leibniz and others pointed out to Huygens, his theories were running counter to the research tradition which they claimed to instantiate. This was an acute conceptual problem of the first magnitude, as Huygens himself sometimes acknowledged. Similarly, when Thomas Young—working within the Newtonian optical research tradition—found himself offering explanations for optical interference which presupposed a wave-theoretic interpretation of light, he was chastised for not recognizing the extent to which his wave theory violated certain canons of the research tradition to which he seemingly paid allegiance.[27] Here again, we can see how the dissonance between a research tradition and its component theories can generate acute conceptual problems.

The constraining role of research traditions. As we have already said, it is the primary function of a research tradition to establish a general ontology and methodology for tackling all the problems of a given domain, or set of domains. As such, it acts negatively as a *constraint* on the types of theories which can be developed within the domain. If the ontology of the research tradition denies the existence of forces acting-at-a-distance, then it clearly rules out as unacceptable any specific theory which relies on noncontact action. It was precisely for this reason that "Cartesians" such as Huygens and Leibniz (committed to an ontology of pushes and pulls) found Newton's theory of celestial mechanics so otiose. Einstein's theory of the equivalence of matter and energy excludes from consideration any specific theory which postulates the absolute conservation of mass. The mechanistic tradition in heat theory (with its corollary that heat can be turned into work) precludes the development of theories which assume the materiality of heat, or heat conservation.

There are also many occasions where the *methodology* of a research tradition rules out certain sorts of theories. For instance, any research tradition which has a strongly inductivist or observationalist methodology will regard as inadmissible "specific" theories postulating entities which cannot be observed. Much of the opposition to subtle fluid theories in the eighteenth century and to atomic theories in the nineteenth century was due to the fact that the dominant methodology of the period denied the epistemic and scientific well-foundedness of theories which dealt with "unobservable entities."[28]

In all these cases, the research tradition within which a scientist works precludes him from adopting specific theories which are *incompatible* with the metaphysics or methodology of the tradition.

Thus far, we have focussed attention primarily on the negative manner in which research traditions exclude certain problems and theories. They also have, however, two very positive functions.

The heuristic role of research traditions. Precisely because they postulate certain types of entities and certain methods for investigating the properties of those entities, research traditions can play a vital heuristic role in the construction of specific scientific theories. Not, of course, because theories can in any sense be deduced from research traditions; but rather because research traditions can provide vital clues for theory construction. Consider the case of Benjamin Franklin and his efforts to articulate a theory of static electricity. Franklin was familiar with certain phenomena (particularly, electrification by friction, electroscopes, and the Leyden jar). Working within a research tradition which postulated the existence of electrical matter, Franklin needed a theory which could explain how friction electrifies bodies, how electrical bodies could attract and repel, how electricity could be stored in a condensor, and why certain bodies were conductors and others were insulators. In the early stages of the development of his theory, Franklin came to the view that positive electrification consisted in the accumulation within bodies of an excess amount of this electrical fluid, while negative electrification was caused by a deficiency of this fluid. If these specific theoretical assumptions are linked together with the ontology of his research tradition, an ontology which postulated that electricity was a form of matter and therefore conserved in the same way that ordinary matter was, it became natural to assume that electrical charge must be conserved. This important theoretical insight, subsequently confirmed in Franklin's experiments, emerged as an almost inevitable result of Franklin's thinking about the relations between his emerging theory and its parent research tradition. It did not follow logically from either the early theory itself, nor from the research tradition. It was the juxtaposition of the two that made possible this vital theoretical extension.

A different sort of heuristic role is illustrated by the early history of thermodynamics. When Sadi Carnot set out to develop a theory of steam engines, he sought to do so within the research tradition of the caloric doctrine of heat. Within this tradition, heat was conceived as a material, conserved substance capable of moving between the constituent parts of macroscopic bodies. Carnot, familiar with the work that could be performed by such simple mechanical systems as a water wheel, tried to conceive of heat flow on analogy with the fall of water, with the temperature gradient between input and output corresponding to the top and bottom heights of the waterfall. It is in terms of this analogy that Carnot develops the "proof" of his theory. It is clear that, if Carnot had not conceived of heat as a conserved substance capable of flowing from one point to another without loss of its quantity, he almost certainly would not have enunciated his theory. But that way of conceiving heat was a natural result of the research tradition within which Carnot worked.

One final example may make the point still clearer. When Descartes attempted to develop a theory of light and colors, he had already defined his general research tradition. In brief, it amounted to the assertion that the only properties which bodies can have are those of size, shape, position, and motion. The research tradition did not, indeed could not, specify precisely what sizes, shapes, positions, and motions particular bodies could exhibit. But it did make it clear that any specific physical theory, in optics or elsewhere, would have to deal exclusively with these four parameters. As a result, Descartes knew—when he set out to explain optical refraction, the colors of the rainbow, and the path of light through lens and prisms—that his optical theories would have to be constructed along such lines. So, he sought to explain colors in terms of the shape and rotational velocity of certain particles; he explained refraction in terms of differential velocities of such particles in different media. Moreover, since his general research tradition made it clear that particles of light are exactly like other material bodies, he recognized that he could apply general mechanical theorems (such as the laws of impact and the principle of conservation of motion) to a theoretical analysis of light. Again, none of his theories followed logically from his research tradition; but, in the ways indicated, that research tradition directed the construction of Cartesian theories in a number of subtle and important ways.

In all the cases mentioned thus far, the research tradition functions heuristically to suggest an *initial* theory for some domain. A second important heuristic role for the research tradition, as Lakatos has pointed out, arises when one of its constituent theories requires modification (because of its lack of problem-solving ability). *Any sound research tradition will contain significant guidelines about how its theories can be modified and transformed, so as to improve its problem-solving capacity.*

For instance, when early versions of the kinetic theory of gases were confronted by some serious predictive failures, there was enormous "flexibility" within the research tradition which pointed the way towards natural modifications that might be made. If more degrees of freedom were needed to accommodate seeming energy losses, kineticists could introduct molecular spin or alter their assumptions about molecular elasticities. If gases did not condense in accordance with theoretical predictions, the addition of weak intermolecular attractions could do the job. All these, and many similar "gambits" emerge quite plausibly from regarding matter as possessing a molecular and mechanical composition.[29]

The justificatory role of research traditions. It is one of the important functions of research traditions to *rationalize* or to *justify* theories. Specific theories make many assumptions about nature, assumptions which are generally not justified either within the theory itself or by the data which confirm the theory. These are usually assumptions about basic causal processes and entities, whose existence and operation the specific theories take "as given." When, for instance, Sadi Carnot developed his theory of the steam engine, his working out of that theory presupposed that no heat was lost in performing the work of driving a piston. (That assumption later turned out to be unacceptable, of course; but it is an assumption which is absolutely crucial to Carnot's "proof" of his theory.) Carnot offered no rationale for that assumption, and, quite rightly, felt no need to; the caloricist research tradition, within which he was working, laid it down as a primary postulate that heat was always conserved. Carnot was thus able to presuppose, in developing his theory, certain things about nature which his theory could not itself establish, not even in principle.

A century earlier, when Stephen Hales had developed his theory about the nature of "air" (i.e., gases), he was able to take it almost for granted that gases were composed of mutually repellent particles, and was able to use repulsion to explain such phenomena as elasticity and gaseous mixing. Had Hales been working in research traditions other than the Newtonian one, such an assumption would have been unthinkable, or at least in need of elaborate justification. (At a minimum, his theory would have had to address itself to justifying that assumption.) But, as a Newtonian, Hales could assume, almost without argument, that it was appropriate and legitimate to conceive of gases as swarms of mutually repellent particles. By sanctioning certain assumptions in advance, the research tradition thus frees the scientist working within it from having to justify all of his assumptions, and gives him the time to pursue specific problems of interest. Although critics outside the research tradition may fault a scientist for constructing theories based on such assumptions, the scientist knows that his *primary audience* —fellow researchers within the same tradition—will not find his working assumptions problematic.

Research traditions thus identify for the scientist working within them three classes of assumptions: those which are unproblematic, because justified by the research tradition; those which are forbidden by the research tradition; and, of course, those which, while not forbidden by the research tradition, definitely require a rationale within the theory (for the research tradition itself provides no rationale for them). Among scientists working within any one research tradition, there will be a broad consensus about where any given statement falls as between those three pigeonholes.

Summing up the discussion thus far, we have seen that such research traditions can justify many of the assertions which their theories make; they can serve to stamp certain theories as inadmissible because they are incompatible with the research tradition; they can influence the recognition and weighting of empirical and conceptual problems for their component theories, and they can provide heuristic guidelines for the generation or modification of specific theories.

The Separability of Theories from Research Traditions

Up to now, I have stressed that virtually all theoretical activity takes place within the context of a research tradition, that such traditions constrain, inspire, and serve to justify the theories which are subsumed under them. Without wishing to negate any of that, it is equally important to recognize that there are circumstances in which theories can break away from the research traditions which initially inspired or justified them. Galileo's theory of fall, for instance, has (since the 1650s) been treated separately from the Galilean research tradition; similar things could be said about Pasteur's theory of disease, Maxwell's theory of electromagnetism, Lavoisier's theory of oxidation, and Planck's theory of black-body radiation, to name only a few cases. Indeed, it is just the eventual possibility of separating a theory from a given research tradition which gives the misleading impression that theories exist independently of, and owe nothing to, research traditions.

This process of theory separation is a fascinating one and deserves to be studied in some detail. I shall limit my remarks here to pointing out that the *separation of a theory from its parent research tradition generally takes place only when that theory can be taken over,* either intact or by small-scale modifications in it, *by an alternative research tradition.* Theories rarely can exist on their own, and even when they do it is only for short periods of time. The reasons for this are clear. Theories are never self-authenticating; they invariably make assumptions about the world for which they provide no rationale. Since it is one of the functions of a research tradition to provide just such a rationale for a theory, it is normally the case that a theory is separated from one research tradition only if it can be absorbed (i.e., justified) within another and more successful research tradition.

The doctrines of early thermodynamics, to which we have referred earlier, are a case in point. Originally developed within a caloricist research tradition (based on substantial, nonkinetic theories of heat) by Carnot and Clapeyron, the theory of thermodynamics proved an embarrassment during the late 1840s and 1850s, by which time the research tradition that inspired it had been largely discredited. There was wide agreement that the theory of thermodynamics was worth preserving, but not (many felt) at the price of subscribing to the research tradition which had generated it. At the same time, the kinetic, anti-caloric research tradition was making great strides forward in other domains, but was thought to be weak insofar as it had been unable to equal, within the area of thermodynamics, the successes which its competitor, the caloric tradition, had achieved. It was Rudolf Clausius, writing in the 1850s, who was able to show that the theory of thermodynamics could be developed and rationalized within the kinetic tradition, independently of the caloricist assumption of the conservation of heat. Clausius thereby showed that the theory of thermodynamics was not inexorably wedded to the caloricist research tradition and could be absorbed by the kineticist tradition. In one fell swoop, Clausius thus managed to strengthen the case for both thermodynamics and for kineticism, by removing what had been a serious conceptual problem for them both. In like manner, Newton (as a vehement opponent of the Cartesian research tradition) was able to show that his own research tradition could absorb the Huygensian theory of impact—a theory which had originally been developed squarely within the Cartesian tradition.

The multitude of cases one could cite of this process ought not lead us to underestimate its difficulty. Precisely because a research tradition plays an important justificatory role for its constituent theories, any alternative research tradition which is to play the same role must be sufficiently rich conceptually, and its partisans sufficiently imaginative, to allow it to justify and rationalize theories which *prima facie* are more naturally related to very different metaphysical and methodological traditions. (I shall have more to say later about this process of "theory appropriation," for it is one of the most important ways in which *new* research traditions establish their scientific credentials.)

The Evolution of Research Traditions

Research traditions, as we have seen, are *historical* creatures. They are created and articulated within a particular intellectual milieu, they aid in the generation of specific theories—and like all other historical institutions—they wax and wane. Just as surely as research traditions are born and thrive, so they die, and cease to be seriously regarded as instruments for furthering the progress of science. I shall consider below how research traditions are displaced by other ones, for the aetiology of research tradition "decay" and "putrefaction" is crucial to the processes which must be understood. For now, however, I want to talk about the ways in which important and substantive changes can occur *within* an on-going research tradition. These changes take two distinct forms.

The most obvious way in which a research tradition changes is by *a modification of some of its subordinate, specific theories.* Research traditions are continuously undergoing changes of this type. Researchers in the tradition often discover that there is, within the framework of the tradition, a more effective theory for dealing with some of the phenomena in the domain than they had realized previously. Slight alterations in previous theories, modifications of boundary conditions, revisions of constants of proportionality, minor refinements of terminology, expansion of the classificatory network of a theory to encompass newly discovered processes or entities; these are but a few of the many ways in which the scientist may seek to improve on the problem-solving success of any of the theories within the research tradition. Whenever he discovers a theory which is a significant improvement on its predecessor he drops the latter immediately. Precisely because a scientist's cognitive loyalties are based primarily in the research tradition rather than in any of its specific theories, he generally has no rational vested interest in hanging onto those individual theories. (It is for just this reason that most individual theories have very short life-spans—in many cases amounting to no more than a few months or even weeks.) Because theories change so rapidly, the history of any flourishing research tradition will exhibit *a long succession* of specific theories.

But there is another important way in which research traditions evolve; this second class of changes involves, not the specific theories within the research tradition, but *a change of some of its most basic core elements.* I must discuss this type of transformation in some detail, since there are many philosophers who have denied that research traditions are capable of any significant internal modification. Both Kuhn and Lakatos, for instance, usually suggest that entities such as research traditions have a rigid and unchanging set of doctrines which identify and define them. Any change in those doctrines, it is suggested, produces a *different* research tradition. Since, Lakatos argues, we define a research tradition or research programme in terms of its central doctrines (doctrines which Lakatos argues we make true by fiat or by convention), any change in those central tenets is *de facto* the abandonment of the research tradition which was defined as the set of those tenets.[30] As tempting as this approach is (for, if true, it would make the process of *identifying* research traditions relatively straightforward), I shall be arguing that we must reject it, for it can only obfuscate our effort to get some understanding of the historical processes of science.

If one looks at the great research traditions in the history of scientific thought—Aristotelianism, Cartesianism, Darwinism, Newtonianism, Stahlian chemistry, mechanistic biology, or Freudian psychology, to name only a few—one can see immediately that there is scarcely any interesting set of doctrines which characterizes any one of these research traditions throughout the *whole* of its history. Certain Aristotelians, at times, abandoned the Aristotelian doctrine that motion in a void is impossible. Certain Cartesians, at times, repudiated the Cartesian identification of matter and extension. Certain Newtonians, at times, abandoned the Newtonian demand that all matter has inertial mass. But need it follow that these seeming "renegades" were no longer working within the research tradition to which they earnestly claimed to subscribe? Does Thomas Aquinas cease to be an Aristotelian because he rejects portions of Aristotle's analysis of motion? Does Huygens become a non-Cartesian because he admits the possibility of void spaces? Certain advantages will accrue if we can plausibly answer these questions negatively. To show how that is possible is the task before us.

A research tradition, we have said, is a set of assumptions: assumptions about the basic kinds of entities in the world, assumptions about how those entities interact, assumptions about the proper methods to use for constructing and testing theories about those entities. In the course of their development, research traditions and the theories they sponsor encounter a number of problems; anomalies are discovered; basic conceptual problems arise. In some cases, proponents of a research tradition will find themselves unable, by modifying *specific* theories within the tradition, to eliminate these anomalous and conceptual problems. In such circumstances, it is common for partisans of a research tradition to explore what sorts of (minimal) changes can be made in the deep-level methodology or ontology of that research tradition to eliminate the anomalies and conceptual problems confronting its constituent theories. Sometimes, scientists will find that there is no amount of tinkering with one or another assumption of the research tradition which will eliminate its anomalies and conceptual problems. This becomes strong grounds for abandoning the research tradition (provided there is some alternative in sight). But, perhaps more often, scientists find that by introducing one or two modifications in the core assumptions of the research tradition, they can both solve the outstanding anomalies and conceptual problems *and* preserve the bulk of the assumptions of the research tradition in tact.

In the latter case, it is positively misleading to speak of the creation of a "new" research tradition, for such language conceals from us the crucial conceptual ancestry and similarity which such cases exhibit. We should speak, rather, of *a natural evolution in the research tradition;* an evolution which represents a change, to be sure, but a change that is far from repudiation of a former research tradition and the creation of a new one.[31]

There is much continuity in an evolving research tradition. *From one stage to the next,* there is a preservation of most of the crucial assumptions of the research traditions. Most of the problem-solving techniques and archetypes will be preserved through the evolution. The relative importance of the empirical problems which the research tradition addresses will remain approximately the same. But the emphasis here must be on *relative* continuity between *successive* stages in the evolutionary process. If a research tradition has undergone numerous evolutions in the course of time, there will probably be many discrepancies between the methodology and ontology of its *earliest* and its *latest* formulation. Thus, the Cartesianism of a Bernoulli, writing a century after Descartes' death, is very different from the Cartesianism of the master. The Newtonian research tradition in Michael Faraday's hands is a far cry from that of Newton's first followers. But a finer-grained analysis of the historical evolution of these research traditions will show that there was a continuous intellectual descent from Descartes to Bernoulli, and from Newton to Faraday, and that the Cartesian and Newtonian research traditions, as different as their end-points may look from their beginnings, exhibited enormous continuity in the character of their transformations.[32]

But such an approach leaves itself open to the obvious challenge: if a research tradition can undergo certain deep-level transformations and still remain in some sense the "same" tradition, how do we distinguish change *within* a research tradition from the replacement of one research tradition by another?

A partial answer to the question comes from recognizing that *at any given time* certain elements of a research tradition are more central to, more entrenched within, the research tradition than other elements. It is these more central elements which are taken, at that time, to be most characteristic of the research tradition. To abandon them is indeed to move outside the research tradition, whereas the less central tenets can be modified without repudiation of the research tradition. Like Lakatos, then, I want to suggest that certain elements of a research tradition are sacrosanct, and thus cannot be rejected without repudiation of the tradition itself. But unlike Lakatos, I want to insist that *the set of elements falling in this (unrejectable) class changes through time.* What was taken to characterize the unrejectable core of the Newtonian tradition in eighteenth-century mechanics (e.g., absolute space and time) was no longer regarded as such by mid-nineteenth-century Newtonians. What constituted the essence of the Marxist research tradition in the late nineteenth century is substantially different from the "essence" of Marxism a half century later. Lakatos and Kuhn were right in thinking that a research programme or paradigm always has certain nonrejectable elements associated with it; but they were mistaken in failing to see that the elements constituting this class can shift through time. By relativizing the "essence" of a research tradition with respect to time, we can, I believe, come much closer to capturing the way in which scientists and historians of science actually utilize the concept of a tradition.

Of course, this still leaves unanswered how it is that scientists decide at any given time which elements of a maxi-theory or research tradition are to be treated as "unrejectable" (a problem likewise unanswered by Kuhn and Lakatos). I cannot give a fully satisfactory answer to the question, but some hunches are probably worth exploring. Both Kuhn and Lakatos seem to believe that the decision about which elements of a maxi-theory fall into this privileged class is arbitrary and not governed by rational considerations: on their account, it simply "happens."[33] I am unable to give a full specification of all the factors which influence the selection of the core of a research tradition, but there are clearly dimensions of the choice which are rational. For instance, one of the major factors influencing the entrenchment of any element of a research tradition is *its conceptual well-foundedness.* The core assumptions of any given research tradition are continuously undergoing conceptual scrutiny. Some of those assumptions will, at any given time, be found to be strong, and unproblematic. Others will be regarded as less clear, less well-founded. As new arguments emerge which buttress, or cast doubt on, different elements of the research tradition, the relative degree of entrenchment of the different components will shift. During the evolution of any active research tradition, scientists learn more about the conceptual dependence and autonomy of its various elements; when it can be shown that certain elements, previously regarded as essential to the whole enterprise, can be jettisoned without compromising the problem-solving success of the tradition itself, these elements cease to be a part of the "unrejectable core" of the research tradition. (For instance, after Mach and Frege argued that none of the other elements of the Newtonian tradition required the absoluteness of space and time, these notions moved perceptibly towards the periphery of the Newtonian research tradition.)

Research Traditions and Changes in Worldviews

We have stressed, both here and in the previous chapter, how research traditions and theories can encounter serious cognitive difficulties if they are incompatible with certain broader systems of belief within a given culture. Such incompatibilities constitute conceptual problems which may seriously challenge the acceptability of the theory. But it may equally well happen that *a highly successful research tradition will lead to the abandonment of that worldview which is incompatible with it, and to the elaboration of a new worldview compatible with the research tradition.* Indeed, it is in precisely this manner that many radically new scientific systems eventually come to be "canonized" as part of our collective "common sense." In the seventeenth and eighteenth centuries, for instance, the new research traditions of Descartes and Newton went violently counter to many of the most cherished beliefs of the age on such questions as "man's place in Nature," the history and extent of the cosmos, and, more generally, the nature of physical processes. Everyone at the time acknowledged the existence of these conceptual problems. They were eventually resolved, not by modifying the offending research traditions to bring them in line with more traditional world views, but rather by forging a new world view which could be reconciled with the scientific research traditions. A similar process of re-adjustment occurred in response to the Darwinian and Marxist research traditions in the late nineteenth century; in each case, the core, "nonscientific" beliefs of reflective people were eventually modified to bring them in line with highly successful scientific systems.

But it would be a mistake to assume that worldviews always crumble in the face of new scientific research traditions which challenge them. To the contrary, they often exhibit a remarkable resilience which belies the (positivistic) tendency to dismiss them as mere fluff. The history of science, both recent and distant, is replete with cases where worldviews have not evaporated in the face of scientific theories which challenged them. In our own time, for instance, neither quantum mechanics nor behavioristic psychology have shifted most people's beliefs about the world or themselves. Contrary to quantum mechanics, most people still conceive of the world as being populated by substantial objects, with fixed and precise properties; contrary to behaviorism, most people still find it helpful to talk about the inner, mental states of themselves and others.

Confronted with such examples, one might claim that these research traditions are still new and that older world views predominate only because the newer insights have not yet penetrated the general consciousness. Such a claim may prove to be correct, but before we accept it uncritically, there are certain more striking historical cases that need to be aired. Ever since the seventeenth century, the dominant research traditions within the physical sciences have presupposed that all physical changes are subject to invariable natural laws (either statistical or nonstatistical). Given certain initial conditions, certain consequences would inevitably ensue. Strictly speaking, this claim should be as true of man and other animals as it is of stars, planets, and molecules. Yet in our own time, as much as in the seventeenth century, very few people are prepared to abandon the conviction that human beings (and some of the higher animals) have a degree of undeterminiation in their actions and their thoughts. Virtually all of our social institutions, most of our social and political theory, and the bulk of our moral philosophy is still based on a worldview seemingly incompatible with a law-governed universe. Despite repeated efforts in the last three centuries to explain away this conceptual problem, it is fair to say that this is one important area where the traditional worldview has made very few concessions to the "broader implications" of some highly successful scientific traditions.[34]

It has long been fashionable to imagine that the worldview or "Zeitgeist" of any epoch always plays a purely *conservative* role, suppressing intellectual innovation and encouraging the retention of the scientific *status quo.* Exponents of scientific progress have frequently bemoaned the use of "worldview" considerations which invariably stifle the emergence of new scientific ideas. E. G. Boring spoke for many scientists and philosophers when he insisted that: "Inevitably by definition the *Zeitgeist* favors conventionality . . . [and] works against originality."[35]

This position is bad philosophy and false history. It is philosophically weak insofar as it ignores the fact that there is no reason, in principle, why an entrenched worldview could not provide a more convincing rationale for an innovative theoretical development than for a traditional theory. Boring's claim that a *Zeitgeist* automatically favors traditional theories is thus without cognitive foundation. The view is equally misleading historically. It is well known, for instance, that the *Zeitgeist* of late seventeenth-century England did much to hasten the replacement of the older mechanical philosophy by the newer science of Newton, precisely because Newton's research tradition could be more readily justified within that framework than the mechanistic science of Descartes could. More recently, the emergence of "new" quantum mechanics in the late 1920s found a quick and ready reception among the many intellectuals who were already convinced that the rigid causal categories of classical science were unreliable.

The Integration of Research Traditions

Up to now I have spoken as if research traditions were invariably in competition with one another, suggesting moreover that the resolution of such a conflict comes when one alone among the competing traditions dominates and when its competitors are effectively vanquished. This is often the case. But it would be a serious error to assume that a scientist cannot consistently work in more than one research tradition. If these research traditions are inconsistent in their fundamentals, then the scientist who accepts them both raises serious doubts about his capacity for clear thinking. But there are times when two or more research traditions, far from mutually undermining one another, can be amalgamated, producing a synthesis which is progressive with respect to both the former research traditions. It is the dynamics of such situations which I want to discuss briefly here.

There are basically two ways in which different research traditions can be integrated. In some cases, one research tradition can be grafted onto another, without any serious modification in the presuppositions of either. Thus, in eighteenth-century natural philosophy, many scientists were simultaneously Newtonians and subtle fluid theorists. Their adherence to the research tradition of subtle fluids (which was as much Cartesian as it was Newtonian) led them to postulate imperceptible aetherial fluids in order to explain the phenomena of electricity, magnetism, heat, perception, and a range of other empirical problems. Their Newtonianism, on the other hand, led them to assume that the constituent particles of such fluids interacted (not by contact, as the Cartesians tried to suggest) but rather by means of strong forces of attraction and repulsion, acting-at-a-distance across empty space. The fusion of these two research traditions was to constitute itself a major research tradition, one which Schofield has labelled "materialism."[36] While undermining the presuppositions of neither of its predecessors, the amalgamation suggested important new lines of research, and put scientists in a position to deal with empirical and conceptual problems which neither of the ancestor traditions alone could resolve satisfactorily.

In other cases, however, the amalgamation of two or more research traditions requires the repudiation of some of the fundamental elements of each of the traditions being combined. In these cases, the new research tradition, if successful, requires the abandonment of its predecessors. (It is, incidentally, in just this way that most so-called scientific revolutions take place; not by the articulation of a research tradition whose *ingredients* are revolutionary and new, but rather by the development of a research tradition whose novelty consists in the way in which old ingredients are *combined.)* There are many examples of this process in the history of any discipline, scientific or otherwise. To consider some scientific cases first, eighteenth- and nineteenth-century natural philosophy is replete with such integrations. Roger Boscovich, for instance, set out deliberately to develop a new "system of nature," by picking and choosing from among the assumptions of two incompatible research traditions, Newtonianism and Leibnizianism. Maupertuis attempted something similar. The work of their contemporary, Daniel Bernoulli, illustrates an analogous attempt to forge a compromise between the research traditions of Cartesian and Newtonian physics. In the eighteenth and nineteenth centuries, geological followers of Hutton were hammering out a new tradition which drew on elements of caloricist heat theories and Vulcanist geology. These research traditions could not be preserved intact and, as a result, the Huttonians had to forge what was regarded as a "revolutionary" research tradition which incorporated elements of traditions which had been previously incompatible. Within economics, Karl Marx drew on elements from the idealism of Hegel, the materialism of Feuerbach, and the "capitalism" of Adam Smith and his English followers.

"Nonstandard" Research Traditions

It would be dishonest to leave the topic of research traditions without adding a caveat, although how important it is remains to be seen. We have thus far characterized research traditions as rather ambitious and grandiose entities, replete with ontologies and methodologies. There is no doubt in my mind that many of the best known research traditions in science possess both these characteristics. But there also seem to be traditions and schools in science which, although lacking one or the other (or in some cases both), have nonetheless had a genuine intellectual coherence about them. For instance, the tradition of psychometrics in the early twentieth century seems to have been held together by little more than the conviction that mental phenomena could be mathematically represented. Equally, the tradition of rational mechanics in the eighteenth century seems to have cut horizontally across almost every conceivable metaphysical and methodological tradition and to have drawn together a group of thinkers committed simply to the mathematical analysis of motion and rest. The important tradition of "analytic physics" in early nineteenth century France (including Biot, Fourier, Ampère, and Poisson) seems to have had no common ontology, although its partisans doubtlessly shared a common methodology. In our own time, cybernetics and information theory seem to be "schools" without well-defined ontologies. Whether, on further investigation, it will turn out that these "nonstandard" research traditions do have ontological and methodological elements or whether, failing that, they will behave differently from "richer" research traditions are still unanswered questions. Much research is still needed on these units that are too narrow to be full-blown research traditions but too global to be mere theories.

The Evaluation of Research Traditions

Our focus thus far has been on the temporal dynamics of research traditions. We have learned something about how such traditions evolve, how they interact with their constituent theories and with wider elements of the worldview and the problem situation.

However, I have said nothing yet about how, if at all, it is possible for scientists to make sensible choices between alternative research traditions, nor about how a single tradition can be appraised relative to its acceptability. This is a crucial issue, for until and unless we can articulate workable criteria for choice between the larger units I am calling research traditions, then we have neither a theory of scientific rationality, nor a theory of progressive, cognitive growth.

In the next few pages, I shall be defining some criteria for the evaluation of research traditions, and discussing some of the different contexts in which cognitive evaluations can be made.

Adequacy and Progress

Even though research traditions in themselves entail *no* observable consequences, there are several different ways in which they can be rationally evaluated and thus compared. Two chief modes of appraisal, however, are the most common and the most decisive. One of these modes is synchronic, the other is diachronic and developmental.

We may, to begin with, ask about the (momentary) *adequacy* of a research tradition. We are essentially asking here how effective the *latest* theories within the research tradition are at solving problems. This, in turn, requires us to determine the problem-solving effectiveness of those theories which presently constitute the research tradition (ignoring their predecessors). Since we already discussed how to evaluate the problem-solving effectiveness of individual theories,[37] we need only combine those appraisals to find the adequacy of the broader research tradition.

Alternatively, we may ask about the *progressiveness* of a research tradition. Here our chief concern is to determine whether the research tradition has, in the course of time, increased or decreased the problem-solving effectiveness of its components, and thus its own (momentary) adequacy. This matter is, of course, unavoidably *temporal;* without a knowledge of the history of the research tradition, we can say nothing whatever about its progressiveness. Under this general rubric, there are two subordinate measures which are particularly important:

1. *the general progress of a research tradition*—this is determined by comparing the adequacy of the sets of theories which constitute the oldest and those which constitute the most recent versions of the research tradition;

2. *the rate of progress of a research tradition*—here, the changes in the momentary adequacy of the research tradition during any specified time span are identified.

It is important to note that the general progress and the rate of progress of a research tradition may be widely at odds. For instance, a research tradition may show a high degree of general progress, and yet show a low *rate* of progress, especially in its recent past. Alternatively, a research tradition may have a high rate of progress during its recent past while exhibiting limited general progress.

Likewise, and even more importantly, the appraisals of a research tradition based upon its progressiveness (either general or time-dependent) may be very different from those based on its momentary adequacy. One can conceive of cases, for example, where the adequacy of a research tradition is relatively high and yet it shows no general progress or even a negative rate of progress. (In fact, many actual research traditions have this character.) Alternatively, there are cases (e.g., behavioristic psychology and early quantum theory) where the general progress and the rate of progress of a research tradition are high, but where the momentary adequacy of the tradition is still quite low.

Needless to say, the appraisals will not always point in contrary directions, but the very fact that they can (and sometimes have) emphasizes the need to attend very carefully to the various *contexts* in which cognitive appraisals of research traditions are made. It is that issue which must occupy us next.

The Modalities of Appraisal: Acceptance and Pursuit

Almost all the standard writings on scientific appraisal, whether we look to philosophical or historical discussions of science, have two common features: they assume that there is only *one* cognitively legitimate context in which theories can be appraised; and they assume that this context has to do with determinations of the empirical well-foundedness of scientific theories. Both these assumptions probably need to be abandoned: the first because it is false, the second because it is too limited.

I shall be arguing that a careful examination of scientific practice reveals that there are generally *two* quite different contexts within which theories and research traditions are evaluated.[38] I shall suggest that, within each of these contexts of inquiry, very different sorts of questions are raised about the cognitive credentials of a theory, and that much scientific activity which appears irrational—if we insist on a uni-contextual analysis—can be perceived as highly rational if we allow for the divergent goals of the following two contexts:

The context of acceptance. Beginning with the more familiar of the two, it is clear that scientists often choose *to accept* one among a group of competing theories and research traditions, i.e., *to treat it as if it were true.* Particularly in cases where certain experiments or practical actions are contemplated, this is the operative modality. When, for instance, a research immunologist must prescribe medication for a volunteer in an experiment, when a physicist decides what measuring instrument to use for studying a problem, when a chemist is seeking to synthesize a compound with certain properties; in all these cases, the scientist must commit himself, however tentatively, to the acceptance of one group of theories and research traditions and to the rejection of others.

How can he make a coherent decision? There are a wide range of possible answers here: inductivists will say "choose the theory with the highest degree of confirmation"; or "choose the theory with the highest utility"; falsificationists—if they give any advice at all—will say "choose the theory with the greatest degree of falsifiability." Still others, such as Kuhn, would insist that *no* rational choice can be made.[39] I have already indicated why none of these answers are satisfactory. My own reply to the question, of course, would be, *"choose the theory (or research tradition) with the highest problem-solving adequacy."*

On this view, the rationale for accepting or rejecting any theory is thus fundamentally based on the idea of problem-solving *progress.* If one research tradition has solved more important problems than its rivals, then accepting that tradition is rational precisely to the degree that we are aiming to "progress," i.e., to maximize the scope of solved problems. In other words, *the choice of one tradition over its rivals is a progressive (and thus a rational) choice precisely to the extent that the chosen tradition is a better problem solver than its rivals.*

This way of appraising research traditions has three distinct advantages over previous models of evaluation: (1) it is *workable:* unlike both inductivist and falsificationist models, the basic evaluation measures seem (at least in principle) to pose fewer difficulties; (2) it simultaneously offers an account of rational *acceptance* and of scientific *progress* which shows the two to be linked together in ways not explained by previous models; and (3) it comes closer to being widely applicable to the actual history of science than alternative models have been.

The context of pursuit. Even if we had an adequate account of theory choice within the context of acceptance, however, we would still be very far from possessing a full account of rational appraisal. The reason for this is that there are many important situations where scientists evaluate competing theories by criteria which have nothing directly to do with the acceptability or "warranted assertibility" of the theories in question.

The actual occurrence of such situations has often been observed. Paul Feyerabend, in particular, has identified many historical cases where scientists have investigated and pursued theories or research traditions which were patently less acceptable, less worthy of belief, than their rivals. Indeed, *the emergence of virtually every new research tradition occurs under just such circumstances.* Whether we look to Copernicanism, the early stages of the mechanical philosophy, the atomic theory in the first half of the nineteenth century, early psychoanalytic theory, the preliminary efforts at the quantum mechanical approach to molecular structure, we see the same pattern: scientists often begin to pursue and to explore a new research tradition long before its problem-solving success (or its inductive support, or its degree of falsifiability, or its novel predictions) qualifies it to be accepted over its older, more successful rivals.

Another side to the same coin is the historical fact that *a scientist can often be working alternately in two different, and even mutually inconsistent, research traditions.* Particularly during periods of "scientific revolution," it is commonly the case that a scientist will spend part of his time working on the dominant research tradition and a part of his time working on one or more of its less successful, less fully developed rivals. If we take the view that it is rational to work with and explore only the theories one accepts (and its corollary that one ought not accept or believe mutually inconsistent theories) then there can be no way of making sense of this common phenomenon.

Hence neither the use of mutually inconsistent theories nor the investigation of less successful theories—both well-attested historical phenomena—can be explained if we insist that the context of acceptance exhausts scientific rationality. Confronted by such cases, we would have to conclude, with Feyerabend and Kuhn,[40] that the history of science is largely irrational. But if, on the other hand, we realize that *scientists can have good reasons for working on theories that they would not accept,* then this frequent phenomenon may be more comprehensible.

To see what could count as "good reasons" here, we must return to some earlier discussions. It has often been suggested in this essay that the solution of a maximum number of empirical problems, and the generation of a minimum number of conceptual problems and anomalies is the central aim of science. We have seen that such a view entails that we should accept at any time those theories or research traditions which have shown themselves to be the most successful problem solvers. But need the *acceptance* of a given research tradition preclude us from exploring and investigating alternatives which are inconsistent with it? Under certain circumstances, the answer to this question is decidedly negative. To see why, we need only consider the following general kind of case: suppose we have two competing research traditions, *RT* and *RT′*; suppose further that the momentary adequacy of *RT* is much higher than that of *RT′*, but that the *rate* of progress of *RT′* is greater than the related value for *RT*. *So far as acceptance is* concerned, *RT* is clearly the only acceptable one of the pair. We may nonetheless decide to work on, further articulate, and explore the problem-solving merits of *RT′*, precisely on the grounds that it has recently shown itself to be capable of generating new solutions to problems at an impressive rate. This is particularly appropriate if *RT′* is a relatively new research tradition. It is common knowledge that most new research traditions bring new analytic and conceptual techniques to bear on the solution of problems. These new techniques constitute (in the cliché) "fresh approaches" which, particularly over the short run, are likely to pay problem-solving dividends. To *accept* a budding research tradition merely because it has had a high rate of progress would, of course, be a mistake; but it would be equally mistaken to refuse to pursue it if it has exhibited a capacity to solve some problems (empirical *or* conceptual) which its older, and generally more acceptable, rivals have failed to solve.

Putting the point generally, we can say that *it is always rational to pursue any research tradition which has a higher rate of progress than its rivals* (even if the former has a lower problem-solving effectiveness). Our specific motives for pursuing such a research tradition could be one of many: we might have a hunch that, with further development, *RT′* could become more successful than *RT;* we might have grave doubts about *RT'* ever becoming generally successful, but feel that some of its more progressive elements could eventually be incorporated within *RT.* Whatever the vagaries of the individual case, if our general aim is increasing the number of problems we can solve, we cannot be accused of inconsistency or irrationality if we pursue (without accepting) some highly progressive research tradition, regardless of its momentary inadequacy (in the sense defined above).

In arguing that the rationality of pursuit is based on relative progress rather than overall success, I am making explicit what has been implicitly described in scientific usage as "promise" or "fecundity." There are numerous cases in the history of science which illustrate the role which an appraisal of promise or progressiveness can have in earning respectability for a research tradition.

The Galilean research tradition, for instance, could not in its early years begin to stack up against its primary competitor, Aristotelianism. Aristotle's research tradition could solve a great many more important empirical problems than Galileo's. Equally, for all the conceptual difficulties of Aristotelianism, it really posed fewer crucial conceptual problems than Galileo's early brand of physical Copernicanism—a fact that tends to be lost sight of in the general euphoria about the scientific revolution. But what Galilean astronomy and physics did have going for it was its impressive ability to explain successfully some well-known phenomena which constituted empirical anomalies for the cosmological tradition of Aristotle and Ptolemy. Galileo could explain, for example, why heavier bodies fell no faster than lighter ones. He could explain the irregularities on the surface of the moon, the moons of Jupiter, the phases of Venus, and the spots on the sun. Although Aristotelian scientists ultimately were able to find solutions for these phenomena (after Galileo drew their attention to them), the explanations proferred by them smacked of the artificial and the contrived. Galileo was taken so seriously by later scientists of the seventeenth century, not because his system as a whole could explain more than its medieval and renaissance predecessors (for it palpably could *not),* but rather because it showed promise by being able, in a short span of time, to offer solutions to problems which constituted anomalies for the other research traditions in the field.

Similarly, Daltonian atomism generated so much interest in the early years of the nineteenth century largely because of its scientific promise, rather than its concrete achievements. At Dalton's time, the dominant chemical research tradition was concerned with elective affinities. Eschewing any attempt to theorize about the microconstituents of matter, elective affinity chemists sought to explain chemical change in terms of the differential tendencies of certain chemical elements to unite with others. That chemical tradition had been enormously successful in correlating and predicting how different chemical substances combine. Dalton's early atomic doctrine could claim nothing like the overall problem-solving success of elective affinity chemistry (this is hardly surprising, for the affinity tradition was a century old by the time of Dalton's *New System of Chemical Philosophy*); still worse, Dalton's system was confronted by numerous serious anomalies.[41] What Dalton was able to do, however, was to predict—as no other chemical system had done before—that chemical substances would combine in certain definite ratios and multiples thereof, no matter how much of the various reagents was present. This phenomenon, summarized by what we now call the laws of definite and multiple proportions, created an immediate stir throughout European science in the decade after Dalton's atomic program was promulgated. Although most scientists refused to accept the Daltonian approach, many nonetheless were prepared to take it seriously, claiming that the serendipity of the Daltonian system made it at least sufficiently promising to be worthy of further development and refinement.

Whether the approach taken here to the problem of "rational pursuit" will eventually prevail is doubtful, for we have only begun to explore some of the complex problems in this area; what I would claim is that the linkage between progress and pursuit outlined above offers us a healthy middle ground between (on the one side) the insistence of Kuhn and the inductivists that the pursuit of alternatives to the dominant paradigm is *never rational* (except in times of crisis) and the anarchistic claim of Feyerabend and Lakatos that the pursuit of *any* research tradition—no matter how regressive it is—*can always be rational.*

Adhocness and the Evolution of Research Traditions

No discussion of the various appraisal vectors utilized in science would be complete without including the notion of adhocness (an issue often discussed under the rubric "independent testability"). At least since the seventeenth century, but particularly in our own era, ad hoc stratagems and hypotheses have received much attention from scientists and philosophers alike.[42] The determination that a theory or theoretical modification is ad hoc gives us grounds, on the usual account, for dismissing it as illegitimate and unscientific. If we are to accept the claims sometimes made by such philosophers as Popper, Grünbaum, and Lakatos,[43] it is irrational or unscientific ever to accept a theory which is ad hoc. What does such adhocness amount to, and why, if at all, is it such a liability for theories which exhibit it?

The issue of adhocness arises most often in connection with the evolution of theories and the manner in which they handle anomalies. We are usually asked to imagine a situation in which some theory, T_1, encounters a refuting instance, A. In response to A, some modification is introduced into T_1, producing T_2. The conventional wisdom insists that the later theory T_2 is ad hoc if: T_2 can solve A, and the other known problems T_1 could solve, but T_2 has no non-trivial, testable implications other than those of T_1 and A. Putting it in the language of this monograph, a theory T_2 is ad hoc if it can solve only those empirical problems solved by its predecessor T_1, and those which constitute refuting instances for T_1, but no further problems.

There are several difficulties with this approach to adhocness. In the first place, we generally have no way of knowing at any given time whether a new theory T_2 will at some later point be able to solve new problems. To make such a judgment sensibly would require a super-human clairvoyance about what empirical problems and what auxiliary theories (which, when conjoined with the theory, might lead to the solution of new problems) are going to emerge in the future. However, taking a cue from Adolf Grünbaum, we can relativize the above definition to situations of belief and say that a theory T_2 is ad hoc if *it is believed* to solve only those empirical problems which were solved by, or refuting instances for, T_1.[44]

But serious difficulties still remain. As Duhem taught us, individual theories *in isolation* generally solve no problems. It is, rather, complexes of theories which are involved in problem solution.[45] Hence, we must modify the traditional characterization once again, yielding a definition such as the following: *a theory is ad hoc if it is believed to figure essentially in the solution of all and only those empirical problems which were solved by, or refuting instances for, an earlier theory.*

Clumsy as it is, this characterization of adhocness seems to do justice to some of the most sophisticated accounts of adhocness developed in the last decade. Assuming that adhocness is understood in this way, we are entitled to ask: *what is objectionable about it?* If some theory T_2 has solved more empirical problems than its predecessor—*even just one more*—then T_2 is clearly preferable to T_1, and, *ceteris paribus,* represents cognitive progress with respect to T_1. However, we can go further than this to claim that the resort to ad hoc stratagems, as defined immediately above, is perfectly consistent with the general aim of increasing our problem-solving capacities. Ad hoc modifications, *by their very definition,* are empirically progressive.

This result should not be surprising. Indeed, much of what we mean by such clichés as "learning from experience" and "the self-correction of science" is represented by situations in which, when a theory encounters an anomaly, we alter the theory so as to transform the anomaly into a solved problem. While it would be a nice bonus if every theory modification could immediately solve some new problems as well as some old, unsolved ones, to insist on that requirement (as, for instance, Popper, Lakatos, and Zahar have) is to repudiate the doctrine that theories which solve more problems about the world are preferable to those which solve fewer.

In urging that adhocness (so defined) is a cognitive virtue rather than a vice, I am clearly not implying that ad hoc theories are invariably better than non-ad hoc ones. My claim, rather, is that an ad hoc theory is preferable to its non-ad hoc predecessor (which was confronted with known anomalies). To believe otherwise is to deny a vital aspect of the problem-solving character of scientific inquiry.[46]

But it might be argued that I have missed the point of the critics of adhocness. They might say, "Yes, of course, T_2 is better than its *refuted* predecessor T_1; but the relevant comparison is between the ad hoc T_2 and some other theory T_n which is not ad hoc but still solves as many problems as T_2." Einstein's special theory of relativity might exemplify T_n while the Lorentz-modified aether theory was T_2.[47] The obvious reply to such criticism is to ask why the admittedly ad hoc character of the Lorentz contraction constitutes a decisive handicap against it in comparing it with special relativity. If the empirical problem-solving capacities of the two theories are, so far as we can tell, equivalent, then they are (empirically) on a par; defenders of the view that the adhocness of T_2 makes it distinctly inferior to T_n must spell out why, in such cases, the comparable problem-solving abilities and equivalent degrees of empirical support can be thrown to the winds simply by stipulating that ad hoc theories are intrinsically otiose.

What seems to lie behind many discussions of adhocness is a conviction—often present but rarely defended—that there is something suspicious about any change in a theory which is motivated by the desire to remove an anomaly. The presumption is that we cannot really trust such cosmetic surgery because, once we know what the anomaly is, it is little more than child's play to produce some face-saving change in the theory which turns the anomaly into a positive instance. I doubt that where "real" science is concerned, this task is such an easy one. We must remember that, as adhocness has been defined, any ad hoc change must *increase* rather than decrease the problem-solving capacity of the theory in question. Most of the obvious and trivial ways of eliminating anomalies—e.g., arbitrarily restricting the boundary conditions, eliminating those postulates of the theory which entailed the anomaly (assuming they could be localized!), redefining terms or correspondence rules—would generally result in *decreasing* the problem-solving effectiveness of a theory. Hence, such maneuvres—which we might well wish to criticize[48]—do not qualify as ad hoc. The detractors of adhocness have yet to show that the emendation of a theory to preserve its problem-solving capacity and to save it from an anomaly requires any less theoretical imagination or serendipity than the construction of a new theory from scratch. To the extent that these same detractors set an epistemic premium on theories which work the first time around, without any juggling or ad hoc adjustments, we are entitled to ask for the rationale for such a preference.

To this philosophical worry, we should briefly add a historical one. Most of the major theories in science—including Newtonian mechanics, Darwinian evolution, Maxwellian electromagnetic theory and Daltonian atomism—were all ad hoc in the sense defined above. Those modern philosophers and scientists who wish to make adhocness a debilitating handicap for any theory which exhibits it must explain why the most "successful" theories of the past were also highly ad hoc.

There is a grain of truth, however, in the worries of many scientists and philosophers about adhocness. To locate it, we must direct our attention away from the empirical level and towards the *conceptual* one. In many of the classic episodes where charges of adhocness have been made (e.g., Ptolemic astronomy, Cartesian physics, phrenology, and the Lorentz-FitzGerald contraction), the cognitive features of the situation can be characterized as follows: a theory, T_1, has encountered an anomaly, A. T_1 has been replaced by T_2, which solves A, and T_1's previous solved problems, but is not known to be able to solve any other empirical problems. At the same time, T_2 has generated more acute conceptual problems than T_1 exhibited (perhaps by making assumptions contrary to the ontology of T_1's research tradition, or by running counter to other acceptable theories). In such cases, the empirical gains made by T_2 may be more than offset by its conceptual losses, resulting in a diminished overall problem-solving effectiveness. Here we would be warranted in refusing to accept T_2 in preference to T_1. Viewed in this light, *the only legitimately pejorative sense of "adhocness" reduces to a situation in which a theory's overall problem-solving effectiveness decreases,* by virtue of its increasing conceptual difficulties. This sort of adhocness is common in science, and is a frequently cited ground for rejecting theories. But it is important to stress that the concept of adhocness itself, thus understood, adds nothing whatever to our analytic machinery for appraising theories, since it is itself just a special case of conceptual problem generation.

I am, by no means, the first to suggest a conceptual interpretation of adhocness; Lakatos, Zahar, and Schaffner have developed similar interpretations recently.[49] In all their discussions, however, conceptual adhocness remains but one of many species of adhocness, rather than the only legitimate sense. Still worse, none of these writers has indicated how conceptual adhocness is to be assessed, nor even what it amounts to. Equally, all these writers leave us in the dark about how seriously, if at all, it should count against a theory if it is ad hoc. The seeming virtue of the approach taken here is that it separates spurious senses of the ad hoc from legitimate ones, and it gives us machinery for assessing the degrees of cognitive threat posed by adhocness to the theories which exhibit it.

Anomalies Revisited

Chapter one contained the paradoxical claim that the refuting instances of a theory are not necessarily anomalous problems, along with a promissory note to clarify that claim once the machinery was available to do so. The evaluational methods outlined here allow us to return to this issue. I said before that a problem was only anomalous (i.e., cognitively threatening) for some theory, *T,* if that problem was unsolved by *T* but solved by one of its competitors. Clearly, some refuting instances will satisfy this definition, but many will not. It is often the case that some prediction of a theory fails to square with the data, but no other available theory can solve the data either. In the latter situation, why should the data not count as a threatening anomaly for *T?*

In brief, the answer is this: Whenever a theory encounters a refuting instance, it is possible to modify the interpretative rules associated with the theory so as to disarm the "refuting" data. If, for instance, we have a theory, *T,* that "all planets move in ellipses" and then discover a satellite of the sun, *S,* which moves in a circle, we can always modify the interpretative rules governing the term "planet" so as to exclude *S*, thus preserving our theory intact and eliminating any appearance of refutation. If there is no other theory extant which can explain the motion of *S*, the exclusion of *S* from *T*'s domain is perfectly reasonable and progressive—for we lose none of our previously won problem-solving successes by legislating *S* out of the relevant domain. By contrast, if some alternative to *T* can solve *S*, then *T*'s legislation of *S* outside the domain is a *regressive* step, open to rational criticism precisely because *T*'s abandonment of *S* as a legitimate problem entails that we sacrifice some of our demonstrated problem-solving capacity.

What this amounts to is that the modification of a theory arbitrarily in order to eliminate a refuting instance is open to criticism only if such a move would lead to a diminished problem-solving efficiency. That can generally be shown to happen only if the refuting instance is solved by some theory in the domain. Hence, a refuting instance only counts as a serious anomaly when it has been solved by some theory or other.

Summary: A General Characterization of Scientific Change

Drawing together the various strands of argument developed in this chapter, we can conclude that:

1. The *adequacy* or *effectiveness* of individual theories is a function of how many significant empirical problems they solve, and how many important anomalies and conceptual problems they generate. The acceptability of such theories is related both to their effectiveness and to the acceptability of their related research tradition.

2. The *acceptability* of a research tradition is determined by the problem-solving effectiveness of its latest theories.

3. The promise, or *rational pursuitability,* of a research tradition is determined by the *progress* (or rate of progress) it has exhibited.

4. Acceptance, rejection, pursuit, and non-pursuit constitute the major cognitive stances which scientists can legitimately take towards research traditions (and their constituent theories).

Determinations of truth and falsity are *irrelevant* to the acceptability or the pursuitability of theories and research traditions.

5. All evaluations of research traditions and theories must be made *within a comparative context.* What matters is not, in some absolute sense, how effective or progressive a tradition or theory is, but, rather, how its effectiveness or progressiveness compares with its competitors.

We can now move on to discuss the implications of this model of scientific progress for an understanding of some of the central historical and philosophical questions about the cognitive growth of science.

Chapter Four

Progress and Revolution

The revolutionary can only regard his revolution as a progress in so far as he is also an historian. COLLINGWOOD (1956), p. 326

The analytic machinery developed in the preceding chapters raises numerous significant questions about the historical evolution and cognitive status of the sciences. The function of this chapter is to examine the ways in which a problem-solving approach to scientific inquiry can throw new light on a number of central historical and philosophical problems about science, and to show how scientific progress, scientific rationality, and the nature of scientific revolutions can all be profitably discussed in terms of the problem-oriented model outlined above.

Progress and Scientific Rationality

One of the thorniest questions of twentieth-century philosophy concerns the nature of rationality. Some philosophers suggest that rationality consists in acting to maximize one's personal utilities; others suggest that rationality consists in believing in, and acting on, only those propositions which we have good grounds for believing to be true (or at least to be more likely than not); others hint that rationality is a function of cost-benefit analysis; still others claim that rationality amounts to no more than putting forward statements which can be refuted. A great deal has been written on these, as well as other, notions of rational belief and rational action. But, ignoring the fact that none of these explications of rationality has been shown to be free of logical and philosophical difficulties, it has *never* been shown that any of them are rich enough to fit our intuitions about the rationality inherent in much of the history of scientific thought. To the contrary, it is relatively easy to show that there are numerous cases in the history of science—cases in which almost everyone would agree intuitively that rational analysis was occurring—which run counter to each of the models of rationality mentioned above.

The theory of research traditions and progress which is outlined in the preceding chapters constitutes a significant improvement on the theories of rationality now in common parlance among philosophers (if by improvement we mean providing a more accurate explication of the cognitive factors present in actual cases of scientific decision making).

As the previous discussion has shown, there are important historical cases where: (1) scientists have invoked what I have called "non-refuting" anomalous problems as major objections to theories; (2) scientists have devoted themselves to the clarification of concepts and to the reduction of other sorts of conceptual problems; (3) scientists have pursued and investigated promising (i.e., highly progressive) theories, even when those theories were less adequate than rivals; (4) scientists have utilized metaphysical and methodological arguments against and in favor of scientific theories and research traditions; (5) scientists have accepted theories confronted by numerous anomalies; (6) the importance of a problem, and even its status as a problem, has exhibited wild fluctuations; (7) scientists have accepted theories which did *not* solve all the empirical problems of their predecessors.

Although cases exhibiting (1) through (7) have not always been rational and cognitively well-founded, the model I have developed allows us to specify circumstances under which *any one of these ploys would be rationally justified.* The same claim cannot be made, I believe, for any other extant model of scientific growth and progress.

But it might well be argued against this model, that it is purely descriptive, with no rational or normative force; that it offers, at best, a taxonomy for identifying certain variables in scientific controversies, but that it does not show why any of those variables *should* play a role in the appraisal of scientific theories. It could be pointed out that *nowhere* do I show how the capacity of a theory to solve problems bears on the truth or the probability of the theory in question. It could be shown that nowhere do I establish that problem-solving ability provides grounds for rational belief.

Some of these criticisms are entirely correct; I do not even believe, let alone seek to prove, that problem-solving ability has any direct connection with truth or probabilities. But I deny that the circumvention of such epistemic questions deprives the model of normative and explanatory force; equally, I deny that a model of rational theory appraisal must issue in judgments of truth, falsity, probability, confirmation, or corroboration.

To make these denials plausible, I must tackle directly, if briefly, the question (so far skirted in this essay) of the connections between rationality and truth.

At its core, rationality—whether we are speaking about rational action or rational belief—consists in doing (or believing) things because we have good reasons for doing so. That does not solve the problem, of course, but only restates it. The restatement, however, is a useful one, for it makes clear that if we are going to determine whether a given action or belief is (or was) rational, we must ask whether there are (or were) sound reasons for it. It is vital to be clear at the outset that many things that would count as good reasons *outside* science cannot constitute good reasons *within* science. To take a trivial example, I might have a good reason for saying that "2 + 2 = 5," if I know that someone will punish me severely if I refuse to say it. Similarly, I might have a good personal reason for trying to resurrect the Ptolemaic theory (if, for instance, I am poor and a research institute of the Vatican begins awarding grants for such projects). But what can count as a good personal reason for doing something does not necessarily count as a good *scientific* reason for doing it. So what does count as a good reason in science? To answer that question, we must consider the aims of science. For if we can show that doing one particular action, rather than another, would be conducive to achieving the aims of the scientific enterprise, then we would have shown the rationality of doing the one and the irrationality of doing the other within the framework of science.

I have tried to argue that the single most general cognitive aim of science is problem solving. I have claimed that the maximization of the empirical problems we can explain and the minimization of the anomalous and conceptual problems we generate in the process are the *raison d'être* of science as a cognitive activity. I have claimed that any research tradition which can exemplify this process through time is a progressive one. It follows from this that *the chief way of being scientifically reasonable or rational is to do whatever we can to maximize the progress of scientific research traditions.* Equally, this line of attack suggests that rationality consists in accepting the best available research traditions. There are, however, other components of rationality which follow from this way of looking at the matter. The model I have outlined suggests, for instance, that scientific debate is rational so long as it involves a discussion of the empirical and conceptual problems which theories and research traditions generate. Thus, contrary to common belief, it can be rational to raise philosophical and religious objections against a particular theory or research tradition, if the latter runs counter to a well-established part of our general *Weltbild*—even if that *Weltbild* is not "scientific" (in the usual sense of the word). The model suggests that the rational appraisal of a theory or research tradition necessarily involves an analysis of the empirical problems which it solves, and the conceptual and anomalous problems which it generates. The model, finally, insists that any appraisal of the rationality of accepting a particular theory or research tradition is trebly relative: it is relative to its contemporaneous competitors, it is relative to prevailing doctrines of theory assessment, and it is relative to the previous theories within the research tradition.

In arguing for this approach to science, I am deliberately driving a wedge between several issues that have hitherto been closely intertwined. Specifically, it has normally been held that any assessment of either rationality or scientific progress is inevitably bound up with the question of the *truth* of scientific theories. Rationality, it is usually argued, amounts to accepting those statements about the world which we have good reason for believing to be true. Progress, in its turn, is usually seen as a successive attainment of the truth by a process of approximation and self-correction. I want to turn the usual view on its head by making rationality parasitic upon progressiveness. *To make rational choices is, on this view, to make choices which are progressive* (i.e., which increase the problem-solving effectiveness of the theories we accept). By thus linking rationality to progressiveness, I am suggesting that we can have a theory of rationality *without presupposing anything about the veracity or verisimilitude of the theories* we judge to be rational or irrational.

If this effort to talk about the cognitive status of scientific knowledge without relating it to the truth claims of such knowledge seems bizarre, one need only consider the circumstances which have motivated this way of tackling the problem. Philosophers and scientists since the time of Parmenides and Plato have been seeking to justify science as a truth-seeking enterprise. Without exception, these efforts have foundered because no one has been able to demonstrate that a system like science, with the methods it has at its disposal, can be guaranteed to reach the "Truth," either in the short or in the long run. *If rationality consists in believing only what we can reasonably presume to be true, and if we define "truth" in its classical, non-pragmatic sense, then science is (and will forever remain) irrational.* Realizing this dilemma, some philosophers (notably Peirce, Popper, and Reichenbach) have sought to link scientific rationality and truth in a different way, by suggesting that although our present theories are neither true nor probable, they are *closer approximations to the truth* than their predecessors. Such an approach offers few consolations, however, since no one has been able even to say what it would mean to be "closer to the truth," let alone to offer criteria for determining how we could assess such proximity.[1] Hence, if scientific progress consists in a series of theories which represent an ever closer approximation to the truth, then science cannot be shown to be progressive. If, on the other hand, we accept the proposal developed in this essay and take the view that science is an inquiry system for the solution of problems, if we take the view that scientific progress consists in the solution of an increasing number of important problems, if we accept the proposal that rationality consists in making choices which will maximize the progress of science, then we may be able to show whether, and if so to what extent, science in general, and the specific sciences in particular, constitute a rational and progressive system.

The price we have to pay for this approach may be regarded by some people as too high, for it entails that we *may* find ourselves endorsing theories as progressive and rational which turn out, ultimately, to be false (assuming, of course, that we could ever definitely establish that any theory was false). But there is no reason for dismay at this conclusion. Most of the past theories of science are already suspected of being false; there is presumably every reason to anticipate that current theories of science will suffer a similar fate. But the presumptive falsity of scientific theories and research traditions does not render science either irrational or non-progressive.

The model under discussion here offers a means of showing how, even granting the fact that every theory of science may well be false, science may nonetheless turn out to be a worthy and intellectually significant enterprise. There will be those who will charge that such an approach is patently instrumentalist and that it entails that science is a hollow set of symbols and sounds, with no bearing on "the real world" or on the "truth." Such an interpretation is very wide of the mark. There is nothing in this model which rules out the possibility that, for all we know, scientific theories are true; equally, it does not preclude the possibility that scientific knowledge through time has moved closer and closer to the truth. Indeed, there is nothing I have said which would rule out a full-bodied, "realistic" interpretation of the scientific enterprise. But what I am suggesting is that we apparently do not have any way of knowing for sure (or even with some confidence) that science is true, or probable, or that it is getting closer to the truth. Such aims are *utopian,* in the literal sense that we can never know whether they are being achieved. To set them up as goals for scientific inquiry may be noble and edifying to those who delight in the frustration of aspiring to that which they can never (know themselves to) attain; but they are not very helpful if our object is to explain how scientific theories are (or should be) evaluated.[2]

The workability of the problem-solving model is its greatest virtue. In principle, we can determine whether a given theory does or does not solve a particular problem. In principle, we can determine whether our theories now solve more important problems than they did a generation or a century ago. If we have had to weaken our notions of rationality and progress in order to achieve this end, we are at least now in a position to be able to *decide* whether science is rational and progressive—a crucial necessity denied to us if we retain the classical connections between progress, rationality, and truth.

How precisely do we go about making this decision? Inevitably, it involves the assessment of specific cases drawn from the history of science; whether science as a whole is rational and progressive depends, of course, upon whether the set of individual choices of theories and research traditions has exhibited progress and rationality. Thus, we may ask whether the reaction of the scientific community to Einstein's paper on the photoelectric effect led to a progressive modification in the theories of physics. At another level, we may ask whether the overall triumph of the Newtonian research tradition over the Cartesian and Leibnizian research traditions in the eighteenth century was progressive. In answering such questions, we must attend very carefully to *the parameters of contemporary scientific debate and controversy,* for it is precisely therein that the historian can find out what the acknowledged empirical and conceptual problems were; it is there that he can get a reasonably clear sense of the weight or importance of those problems. By a subtle analysis of the actual case (and not a so-called rational reconstruction of it), the historian—or the contemporary scientist—can usually determine the degree to which competing research traditions, or competing theories within the same research tradition, were progressive in their modifications.

What is crucial here is that we must cast our nets of appraisal sufficiently widely that we include *all* the cognitively relevant factors which were *actually present* in the historical situation. We must not assume *a priori,* as some historians of science have, that the only important parameters were experimental or other obviously "scientific" ones. Because theories and research traditions have to be accommodated within a broader network of beliefs and preconceptions, any accurate appraisal of an episode must attend carefully to the philosophical, theological and other intellectual currents which were brought to bear on the case at hand. The fact that a twentieth-century scientist might not recognize the cogency of an objection to a theory on philosophical or religious grounds manifestly does *not* mean that an understanding of the rationality of earlier science can be acquired by ignoring such factors. If a culture at a particular time has a strongly entrenched set of religious or philosophical doctrines which thinkers in that culture believe to be crucial to an understanding of nature, then it is perfectly rational to appraise new scientific theories or research traditions in light of their ability to be accommodated within that prior system of beliefs and presuppositions.

There are doubtless those who would argue that such an approach so relativizes our standards of rationality that it will justify any set of beliefs. If such a criticism were true, then there would be grave problems with the notion of rationality being defended here. But that is far from the case. To suggest that "anything goes," that any combination of beliefs would emerge as rational and progressive on this model, is profoundly *to misunderstand the high standards of rational behavior which it requires.* Nor does the model involve the complete surrender of our standards of rationality to the exigencies of earlier times and places.

This point is worth discussing at some length, for it bears crucially on many of the central dilemmas in the historiography and sociology of science. Many philosophers have sought to set up standards of rationality or progress which are valid for all times and places. They see the task of the historian-philosopher of science as that of evaluating historical episodes entirely with respect to *modern* theories of rational acceptance and appraisal.

In some cases, proponents of such an approach have gone so far as to claim that all the actual standards of rational appraisal *have remained constant through time.* Israel Scheffler, for instance, summarizes this view as follows:

> Underlying historical changes of theory, . . . [is] a constancy of logic and method, which unifies each scientific age with that which preceded it. . . . Such constancy comprises not merely the canons of formal deduction, but also those criteria by which hypotheses are confronted with the test of experience and subjected to comparative evaluation.[3]

We need waste little time on this approach. Virtually all the scholarly literature on the history of methodology shows unambiguously that such components of rational appraisal as criteria of explanation, views about scientific testing, beliefs about the methods of inductive inference and the like have undergone enormous transformations.

A second group, represented by Popper and Lakatos, acknowledges that scientific standards of rationality have evolved, but insists that we should evaluate historical episodes using *our* standards and simply ignore the appraisals made by the relevant scientists about the rationality of what they were doing. On this approach, we pay no heed to whether an experiment was viewed as reliable, whether a theory was regarded as intelligible, or whether an argument was perceived as cogent.[4] What matters, rather, is whether, *by our lights,* a particular theory was well-founded.

Understandably, historians have been dismayed by both these approaches. What is the point, they ask, of analyzing the rationality of past science unless we take into account the views of historical agents about the rationality of what they were doing? Unencumbered by modern notions of rationality, scientists of the past had to make decisions about the acceptability of contemporary theories by *their* criteria rather than by *ours.* We may have the hubris to imagine that our theories of rationality are better than theirs (and they may well be), but how does it help *historical* understanding to evaluate the cogency of past theories utilizing evaluative measures which we know were *not* operative (not even in an approximative form) in the case at hand?

But the historian is confronted by the other horn of the dilemma. If he simply takes at face value every actual appraisal by past scientists of the rationality of a belief, he will never be in any position to judge whether such appraisals were, even by the appropriate standards of the time, well-founded. Obviously, the fact that some historical agent says, "theory A is better than theory B," does not necessarily make it so. If the historian is to explain why certain theories triumphed and others perished, then he must (unless he takes the view that theory choice is always irrational) be able to show that some theories—by the best available rational standards of the time—were superior to others.

Hence, the central problem seems to be this: how can we, with the philosophers, continue to talk normatively about the rationality (and irrationality) of theory choices in the past, while at the same time avoid the grafting of anachronistic criteria of rationality onto those episodes?

The model I have outlined resolves part of that difficulty by exploiting the insights of our own time about the *general* nature of rationality, while making allowances for the fact that many of the *specific* parameters which constitute rationality are time- and culture-dependent. It transcends the particularities of the past by insisting that for all times and for all cultures, provided those cultures have a tradition of critical discussion (without which no culture can lay claim to rationality), rationality consists in accepting those research traditions which are most effective problem solvers. It insists that for scientists in any culture to espouse a research tradition or a theory which is less adequate than other ones available *within* that culture is to behave irrationally. In these important respects, the model argues that there are certain very general characteristics of a theory of rationality which are *trans-temporal* and *trans-cultural,* which are as applicable to pre-Socratic thought, or the development of ideas in the Middle Ages, as they are to the more recent history of science. On the other hand, the model also insists that what is specifically rational in the past is partly a function of time and place and context. The kinds of things which count as empirical problems, the sorts of objections that are recognized as conceptual problems, the criteria of intelligibility, the standards for experimental control, the importance or weight assigned to problems, are all a function of the methodological-normative beliefs of a particular community of thinkers. The model under discussion here possesses the advantage of allowing us to integrate the specific historical norms of a previous epoch and the more general, time-independent features of rational decision making.[5]

To ignore the time-specific parameters of rational choice is to put the historian or philosopher in the outrageous position of indicting as irrational some of the major achievements in the history of ideas. Aristotle was not being irrational when he claimed, in the fourth century B.C., that the science of physics should be subordinate to, and legitimated by, metaphysics—even if that same doctrine, at other times and places, might well be characterized as irrational. Thomas Aquinas or Robert Grosseteste were not merely stupid or prejudiced when they espoused the belief that science must be compatible with religious beliefs.

We in the twentieth century may vehemently disagree with such views, thinking them obscurantist and harmful to the development of science. And in so disagreeing, I believe we are right. One of the things that time has shown is that theories and research traditions sometimes (though not always) flourish best when they are not subordinated to the theological and metaphysical doctrines dominant outside the scientific community. But it is with the advantage of hindsight that we have come to that conclusion. In the absence of the experience of the last three centuries, it would be palpably absurd to assume that it was irrational to imagine that science, theology and metaphysics could be mutually supportive. *The view that science is quasi-independent of such disciplines is itself a research tradition,* one of relatively recent origin. It is a kind of research tradition that has, in its way, generated a considerable degree of progress. That is why it may be rational in the twentieth century to accept it. But the fact that a belief is rational in the present age, or in any age for that matter, does not necessarily entail that it was rational at other times and places. Quite the reverse is more often the case.

It should be clear by now that, in arguing that the cultural exigencies and pressures exerted on science must be taken into account, I am neither abandoning the possibility of rational appraisal nor am I insisting that nonscientific factors are present in every case of scientific choice. I am simply suggesting that we need a broadened notion of rationality which will show how the "intrusion" of seemingly "nonscientific" factors into scientific decision making is, or can be, an entirely rational process. Far from viewing the introduction of philosophical, religious and moral issues into science as the triumph of prejudice, superstition and irrationality, this model claims that the presence of such elements may be entirely rational; further, that the suppression of such elements may itself be irrational and prejudicial.

Of course, whether it is rational to use theological, moral, or philosophical arguments for (or against) a new scientific theory or research tradition is a contingent matter which depends on how rational and progressive are the research traditions which provide such arguments. To argue against modern theories of chemical combustion on the grounds that such theories are incompatible with the myth of Vulcan is patently absurd, for the Greek myths have scarcely established themselves as a body of rational and progressive dogma. To argue against Marxist economics on the grounds that it is contrary to Christian morality is, again, to use a singularly non-progressive tradition as a tool for criticizing a relatively progressive "scientific" tradition. The rationality or irrationality of any episode where "nonscientific," but intellectual, factors play a role must be assessed on a case-by-case basis. But the guiding principles here should be these: (1) in the case of competing scientific research traditions, if one of those traditions is compatible with the most progressive "worldview" available, and the other is not, then there are strong grounds for preferring the former; (2) if both traditions can be legitimated with reference to the same worldview, then the rational decision between them may be made on entirely "scientific" grounds; (3) if neither tradition is compatible with a progressive worldview, their proponents should either articulate a new, progressive worldview which does justify them, or develop a new research tradition which can be made compatible with the most progressive extant worldview.

Scientific Revolutions

For well over a century, it has been commonplace to focus on "scientific revolutions" as one of the core concepts for historical narration and exegesis. Within the last two decades, the idea of a revolution has become canonized in Thomas Kuhn's classic *The Structure of Scientific Revolutions.* Although far from his intent (since Kuhn was primarily concerned to draw attention to non-revolutionary, "normal science"), his book has led many scientists, philosophers and historians alike to compartmentalize the evolution of science into widely separated periods of revolutionary activity, and to imagine that the scientific revolution (with its attendant "change of paradigm") is the basic category for discussing the evolution of science.

Although scientific revolutions are undoubtedly important historical phenomena, they possess neither the importance nor the cognitive character often associated with them. They have assumed this privileged position largely because their structure has been mis-described in ways that make them seem radically unlike science in its usual state; the exaggeration of the difference between "normal" and "revolutionary science" in its turn has led some writers to lay heavier stress on "periods of revolutionary activity" than they probably deserve.

Consider, for instance, Kuhn's account of scientific revolutions. For him, a revolution is marked by the emergence of a new theoretical "paradigm" which, in a short span of time, discredits the older paradigm and draws the virtually unanimous adherence of every member of the relevant scientific community. Revolutions, on his view, are preceded by short periods of frenetic theoretical activity during which many alternative viewpoints vie for the allegiance of the scientific community. Elements of the previous paradigm which were previously sacrosanct suddenly become objects of lively debate and heated controversy. A wide range of alternative viewpoints is explored until eventually (usually in less than a generation) one of these new views vanquishes all the others and becomes established as the new paradigm, demanding unquestioning adherence from scientists in the field. Indeed, Kuhn even goes so far as to say that a discipline is *not scientific* if the discussion of critical, foundational problems continues unabated.[6] If revolutions really had such a character, if they really differed so much from "normal science," they would, of course, be singularly interesting historical phenomena (both from a conceptual *and* from a sociological point of view).

There is much evidence to suggest, however, that scientific revolutions are not so revolutionary and normal science not so normal as Kuhn's analysis would suggest. As we have already observed, debate about the conceptual foundations of any paradigm or research tradition is a historically continuous process. The posing and resolving of conceptual problems—a phenomenon Kuhn relegates chiefly to short-lived periods of crisis—continues unabated throughout the life of any active research tradition. As several critics have noted, Kuhn and his followers have been unable to point to any lengthy period in the history of any major paradigm when its partisans closed their eyes to the conceptual problems which the paradigm generated. One important reason why these basic framework questions rarely go away springs from another feature of science which Kuhn has ignored; namely, the rarity with which any one paradigm achieves that hegemony in the field which Kuhn requires for "normal science." Whether we look at nineteenth-century chemistry, eighteenth-century mechanics, twentieth-century quantum theory; whether we examine evolutionary theory in biology, mineralogy in geology, resonance theory in chemistry or proof theory in mathematics, we see a far more diversified situation than Kuhn's account allows. Two (or more) research traditions in each of these areas have been the rule rather than the exception. Indeed it is difficult to find any extended period of time (even on the order of a decade) when only one research tradition or paradigm stood alone in any branch of science.

It may be helpful to select some of Kuhn's own examples to see how badly his analysis founders:

The Newtonian revolution in mechanics. Like several other scholars, Kuhn's archetypal example of a scientific revolution is the development of Newtonian mechanics from 1700 to the middle of the nineteenth century; this is scarcely surprising since there can have been few more successful paradigms or research traditions than this one. But eighteenth-century mechanics offers few consolations for a Kuhnian theory of revolutions. From its first reception at the hands of Huygens and Leibniz, its core assumptions were under continuous critical scrutiny, even from many physicists who readily conceded its mathematical virtuosity and its empirical triumphs.[7] George Berkeley, several of the early Bernoullis, Maupertuis, the Hutchinsonians, Boscovich, the young Kant, and even Euler raised a number of fundamental problems about the *ontological* foundations of Newtonian mechanics. At the same time, many other scientists (e.g., Hartley, LeSage, Lambert) were taking issue with the *methodological* assumptions of the Newtonian tradition.[8] Although there can be no doubt that the Newtonian tradition had a tremendous impact on eighteenth-century rational mechanics, that tradition exhibited neither the unanimity of adherence nor the suspension of critical judgment which, on Kuhn's view, typify the aftermath of a scientific revolution.

The Lyellian revolution in geology. On Kuhn's account, it was the publication of Charles Lyell's *Principles of Geology* (1830-33) which established the first important scientific tradition in geology.[9] In other words, Lyell's *Principles* provided both a paradigm ("uniformitarianism") and some working exemplars for geology, which collectively constituted a scientific revolution. Even on the most charitable interpretation of the historical evidence, the Lyellian revolution does not support Kuhn's historiography. In the first place, there was nothing *global* about the Lyellian revolution. Restricted largely to England and America, Lyell's work was hardly ever taken seriously in Germany and France, and virtually no continental geologist became a "Lyellian." Even within the English-speaking world, Lyell's ideas—though widely cited—were severely criticized and rarely accepted without emendation. Indeed, the most characteristic features of Lyell's geological system (namely, his "degree uniformitarianism," his theory of climate, and his volcanic theories) were accepted by very few geologists. Equally, there was none of that *cessation of foundational debate* associated with the end of a Kuhnian revolution. In the two generations after Lyell's work most geologists, cosmogonists, geographers, and bio-geologists (most notably Charles Darwin) found it necessary to abandon many of the most fundamental assumptions of the Lyellian paradigm (e.g., Lyell's conviction that the entire spectrum of animals and plants is fully represented in each geological epoch). Even before evolutionary theory discredited Lyellian geology, many critical voices had been raised against virtually all of its core presuppositions. What is true of Lyell is equally applicable to the whole of early nineteenth century geology: there was no geological paradigm which was either universally or uncritically accepted. A multiplicity of alternative frameworks was the rule rather than the exception.

It is this *perennial co-existence of conflicting traditions of research* which makes the focus on revolutionary epochs so misleading. These traditions are constantly evolving, their relative fortunes may shift through time, old traditions may be largely displaced by new ones, but it is generally unhelpful to focus attention on certain stages of this process as revolutionary and on others as evolutionary. The examination of fundamentals, the exploration of alternative frameworks, the replacement of older perspectives by newer and more progressive ones take place unceasingly in science—and in every other intellectual discipline for that matter. This is not to say, of course, that every scientist is (as Popper would have it) constantly criticizing the framework or tradition within which he works. Many scientists at any given time will be taking the tradition as "given" and will be seeking constructively to apply it to a wider range of unsolved empirical problems (what Kuhn calls "puzzle solving"). But to imagine that all scientists are doing that all of the time—except in rare periods of crisis—is to take remarkable liberties with the actual evolution of the sciences.

Clearly, if the notion of a scientific revolution is to be

historically fruitful, we must be able to define scientific revolutions in such a way that their occurrence allows for a persistent disharmony among scientists concerning the basic foundations of their discipline.

One natural approach here would seem to involve a discussion of numbers. One might suggest, for instance, that a scientific revolution occurs when a sizable number of influential scientists in any discipline abandon one research tradition and espouse another. But what constitutes a "sizable number"? This is not a mere matter of counting heads, or speaking of a revolution occurring as soon as more than half of the scientific community adopts one particular research tradition. *Revolutions can be,* and often have been, *achieved by a relatively small proportion of scientists in any particular field.* Thus, we speak of the Darwinian revolution in nineteenth-century biology, even though it is almost certainly the case that only a small fraction of working biologists in the last half of the nineteenth century were Darwinians. We speak of a Newtonian revolution in early eighteenth-century physics, even though most natural philosophers in the period were not Newtonians. As we have seen, it is common to speak of Lyell as having wrought a revolution in geology, even though the bulk of his scientific contemporaries had grave reservations about the research tradition which he espoused.

Examples such as these suggest that a scientific revolution occurs, not necessarily when all, or even a majority, of the scientific community accepts a new research tradition, but rather when a new research tradition comes along which generates enough interest (perhaps through a high initial rate of progress) that scientists in the relevant field feel, whatever their own research tradition commitments, that they have to come to terms with the budding research tradition. Newton created the stir he did because, once the *Principia* and the *Opticks* were published, almost every working physicist felt that he had to deal with the Newtonian view of the world. For many, this meant finding cogent arguments *against* the Newtonian system. But what was almost universally agreed was that Newton had developed a way of approaching natural phenomena which could not be ignored. Similarly, late nineteenth-century biologists, whether fervent Darwinians or confirmed anti-evolutionists, found themselves having to debate the merits of Darwinism. To put the point in a more general fashion, I am suggesting that *a scientific revolution occurs when a research tradition, hitherto unknown to, or ignored by, scientists in a given field, reaches a point of development where scientists in the field feel obliged to consider it seriously as a contender for the allegiance of themselves or their colleagues.*

It is worth noticing that I have defined revolutions in such a way as to presuppose nothing whatever about their inherent rationality or progressiveness. Scientific revolutions can occur even when it is entirely irrational or nonrational considerations which bring a new research tradition to everyone's attention. A revolution could, in principle, involve the abandonment of more progressive research traditions for less progressive ones. In short, whether a scientific revolution is rational and progressive is a *contingent* matter. In sharp contrast to Kuhn, who argues that scientific revolutions are *ipso facto* progressive,[10] I want clearly to separate the question of whether a revolution has occurred from a determination of the progressiveness of that revolution. Otherwise, the claim that science is progressive becomes vacuously true, and therefore cognitively worthless.

Even so conceived, it must still be stressed that scientific revolutions are not the core unit for analysis which some historians and philosophers have imagined. Once we accept that the emergence of new research traditions, and the criticism and modification of older ones *is* the "normal" state of science, then a preoccupation with revolutions—as historical phenomena different in kind from ordinary science—must be avoided. But we can go further than this. If theories and research traditions are undergoing *continuous* appraisal and evaluation, then the natural focus for the historian's interest should be specific research traditions and the debates about the relative merits of the extant traditions in any science. A successful revolution is nothing more than a consequence of, an obituary for, a particularly dramatic and decisive encounter between vying research traditions.

Revolution, Continuity, and Commensurability

Among scholars who have thought about the process of scientific change, there is a central cleavage between those who are struck by the successive revolutionary convulsions of scientific thought, and those who are more vividly impressed by the remarkable continuities which science exhibits through its history. The "revolutionary" school stresses the very different kinds of metaphysics of nature implicit in successive scientific periods. Thus, Aristotle believed in a plenum; seventeenth century atomists, in a void. Eighteenth-century chemists believed that air was composed of highly reactive chemical substances and that fire was not an element. Seventeenth-century and eighteenth-century geologists saw the history of the earth in terms of processes of change and transformation quite unlike anything still occurring today on the face of the globe; some nineteenth-century geologists, on the other hand, were impressed by the uniformitarian character of the earth's history.

By contrast, the "gradualists" stress the degree to which science manages to preserve most of what it has discovered. They point out that, for all the seeming "revolutions" in optics since the early seventeenth century, we still espouse substantially the same sine law of refraction that Descartes did. They point out that, Einstein notwithstanding, contemporary mechanics still utilizes almost entirely techniques worked out by Newtonian scientists, or reasonable approximations thereto. The gradualists see the process of knowledge acquisition as slow and cumulative, with new truths, or better approximations being constantly added to a reservoir of laws about nature which have accumulated since antiquity. They point out, further, that many seemingly radical conceptual innovations often amount to little more than a subtle juxtaposition or realignment of traditional elements.

Both historiographical schools have focussed on important traits of the history of science, but neither has managed to integrate them convincingly. When regarded from the point of view of a problem-solving approach, it is easier to capture both sorts of insights. The chief element of continuity, one might say, is the base of empirical problems. Although there is some change in the empirical problem domain as a result of both time and successive research traditions, what continuity there is in science tends to be found in the domain of such empirical problems. Since the 1640s, every optical theory has had to deal with what happens to light when it is refracted through a prism. Since antiquity, every astronomical theory has been obliged to explain solar and lunar eclipses. Since the 1650s, every theory of matter and of the gaseous state has been forced to explain the (approximately) inverse relations between pressure and volume of gases. Since about 1800, every theory of chemistry has had to grapple with the role of air in the process of combustion. History suggests that problems such as these are a *permanent* fixture of the scientific scene and that, however much the basic ontology of science changes, however many new research traditions emerge, many of these problems will be essential explananda for science throughout its evolution.

Where discontinuities occur is not so much at the level of first order problems as at the level of explanation or problem solution. There are radical differences between the way in which a contemporary chemist explains combustion and the manner in which his eighteenth- or nineteenth-century predecessors did. There are crucial discontinuities between the quantum physicist's explanation of black-body radiation and a nineteenth-century physicist's account of the same problem. Of course, this is not to suggest that successive research traditions have nothing in common except a partial overlap of their empirical problems. There are often important formal and conceptual relationships which persist through time, and which are preserved in a succession of research traditions. But *it is basically the shared empirical problems which establish the important connections between successive research traditions;* it is these, and these alone, which must be preserved if science is to exhibit that (partially) cumulative character which is so striking about much of its history.[11]

Many recent writers on the problem of scientific change, particularly those in the "revolutionary" camp, have been struck by the radical *incommensurability* between successive research traditions. Carrying the "revolutionary" stance to its extreme, they argue that theories before and after a revolution are so radically different that we cannot even speak meaningfully of any similarities between them. Quite rightly pointing out that a Ptolemaian and a Copernican, or a Lamarckian and a Darwinian, or a Newtonian and a relativist look at the world in different ways (perhaps even "look at" different worlds, although that strikes me as a bizarre way of stating the issue), these writers (such as Hanson, Quine, Kuhn, and Feyerabend) have drawn some very pessimistic conclusions about the possibility of rationality in science. In several cases, they have been led to conclude that it is, *in principle,* impossible to establish that any research tradition ever rationally triumphs over another. The logic of their arguments (which I shall examine shortly) leads them to conclude that the history of science is but a *succession* of different worldviews, and that rational choice can never be made between such divergent schemes of the universe. Because each has its own internal rationale and integrity, no meaning can be attached to the suggestion that one scheme is more (or less) rational than another.

This argument is an important one. If true, it would mean that science had no particular claim to our cognitive loyalities. If there are no conceivable grounds for rational choice between competing research traditions, then science becomes a matter of whim and caprice, in which that tradition wins which happens to attract the most influential adherents and the most powerful propagandists. That may well be how science *is,* but before we accept the rather depressing conclusion that science must necessarily be that way, it is worth examining with some care the arguments which its proponents give for such a relativistic notion of scientific change.

In brief, the central argument runs like this: scientific theories implicitly define the terms which occur within them. Hence, if two theories are different then all the terms within them must have different meanings. (Thus, when an Einsteinean physicist refers to the "mass" of a particle, he means something different from a Newtonian when the latter refers to the "mass" of a particle.) Moreover, the argument continues, even the so-called observational reports which scientists working with different theories make are incommensurable, for their observational terms are theory-laden, i.e., they are given meaning by virtue of one or another theory. That means that, although scientists working in different research traditions may sometimes make the same verbal utterances, we cannot even then assume that they are asserting the same thing. To accept a theory, on this view, is to accept a quasi-private language which no one who does not accept the same theory can understand or comprehend. As a result, scientists working in different research traditions cannot communicate with, and cannot understand the statements of, their fellow scientists in other traditions. Given this general incomprehension, science emerging as a new version of the tower of Babel, theories cannot be compared and rationally evaluated because such comparison seemingly requires a common language for speaking about the world.

I believe this general argument to be faulty in several respects. It rests on a very idiosyncratic theory about how words acquire meaning (namely, the theory of implicit definition).[12] It begs a number of questions about synonymy and translation. But *its central flaw,* for our purposes, *lies in its presumption that rational choice can be made between theories only if those theories can be translated into one another's language or into a third "theory-neutral"* language. As Kuhn puts the point, "[the] comparison of two successive theories demands a language into which at least the empirical consequences of both can be translated without loss or change."[13] I shall maintain, to the contrary, that even if we accept the view that all observations are theory-laden to a degree that makes their contents inseparable from the theory which is used to express them, it is still possible to outline machinery for objective, rational comparisons between competing scientific theories and research traditions. There are two general arguments which lead me to such a conclusion.

The argument from problem solving. In the heyday of logical positivism, it was commonly argued that competing theories could be evaluated by comparing their "observational" consequences. Given the dominance at that time of the linguistic metaphor, this was usually conceived as a process of translating the predictions of competing theories (via so-called correspondence rules) into some purely observational language. Because the observational language was held to be free of any speculative, theoretical biases, it was thought to provide objective grounds for the empirical appraisal of vying theories. As doubts grew about the existence of rules of correspondence and about the existence of theory-free observation languages, philosophers such as Kuhn, Hanson, and Feyerabend began to despair about the possibility of any objective yardstick for comparing different theories and suggested that theories were incommensurable and thus not open to objective comparison.

What this approach ignores is that *neither* correspondence rules *nor* a theory-free observation language are necessary for comparing the empirical consequences of competing theories. For even *without* correspondence rules and *without* a purely observational language, we can still talk meaningfully about different theories being *about the same problem,* even when the specific characterization of that problem is crucially dependent upon many theoretical assumptions.

How, acknowledging the manner in which theories affect our characterization of what problems are, can we nonetheless show that different theories address the "same" problem? The answer is straightforward: the terms in which a problem is characterized will generally depend upon the acceptance *of a range of theoretical assumptions,* $T_1, T_2, \ldots, T_3$. These assumptions may, or may not, constitute the theories which solve the problem. If a problem can be characterized only within the language and the framework of a theory which purports to solve it, then clearly no competing theory could be said to solve the same problem. However, *so long as the theoretical assumptions necessary to characterize the problem are different from the theories which attempt to solve it, then it is possible to show that the competing explanatory theories are addressing themselves to the same problem.* Consider a very elementary example. Since antiquity, scientists have been concerned to explain why light is reflected off a mirror or other polished surface according to a regular pattern. Relating the incident to the reflected angle, the problem of reflection, thus characterized, involves many quasi-theoretical assumptions, e.g., that light moves in straight lines, that certain obstacles can change the direction of a ray of light, that visible light does not continuously fill every medium, etc. Does the existence of these theoretical assumptions entail that no two theories can be said to solve the problem of reflection? The answer is clearly negative, provided that the theories which solve the problem are not inconsistent with those relatively low-level theoretical assumptions required to state the problem.[14] Throughout the late seventeenth century, for example, numerous conflicting theories of light (including those of Descartes, Hobbes, Hooke, Barrow, Newton, and Huygens) addressed themselves to the problem of reflection. The various optical theories were all regarded as solving the problem of reflection, because that problem could be characterized in a way which was independent of any of the theories which sought to solve it.

I do not mean to suggest, of course, that *all* the problems which a theory or research tradition attempts to solve can be characterized independently of the theory (or theories) which solves them. The determination of the "independence" of any specific problem must depend upon the particularities of the case. It is my impression, however, that there are far more problems common to competing research traditions than there are problems unique to a single one. These shared problems provide a basis for a rational appraisal of the relative problem-solving effectiveness of competing research traditions.

I must stress again that this argument does *not* presuppose that empirical problems can be stated in some purely observational, nontheoretical language. To speak of (for instance) light being refracted through a prism makes a number of theoretical assumptions (among them, that light moves, that something happens to light while it is "inside" a prism, etc.). It is not the atheoretical nature of empirical problems which is being alleged here. Rather, the weaker claim being made is this: *with respect to any two research traditions (or theories) in any field of science, these are some joint problems which can be formulated so as to presuppose nothing which is syntactically dependent upon the specific research traditions being compared.* Thus, when eighteenth-century Newtonians and Cartesians spoke about the problem of free fall, they were identifying the same problem—for all the profound differences between their respective research traditions. When these same natural philosophers discussed the problem why all the planets moved in the same direction about the sun, they were also in complete agreement about the nature and meaning of the problem (although they did quarrel about its relative *importance* as a problem). When early nineteenth-century geologists debated the explanation of stratification, they could all—whether uniformitarian or catastrophist, whether Neptunist or Vulcanist, whether Huttonian or Wernerian, whether God-fearing or God-denying, whether French, English or German—agree that one problem for any geological theory was that of explaining how such uniform and distinct layers had been formed.

Kuhn has been misled by his discovery that some empirical problems are not jointly shared between different traditions or paradigms (which is certainly true) into believing that *no* problems are identical. The generalized thesis of problem incommensurability is as perverse as the limited thesis of partial non-overlap is profound.

The argument from progress. The argument just given presumes that there are ways of identifying and characterizing some problems which are neutral with respect to the various theories which attempt to solve those problems. But there are doubtless those philosophers who will deny that there is any way in which empirical problems can be characterized so as to allow us to speak of "two theories (or research traditions) solving (or failing to solve) the same problem." I have yet to see any compelling arguments to that effect, but even if there were—i.e., even if we grant that it cannot be decided whether theories are dealing with the same problems—there is still scope for the objective evaluation and comparison of incommensurable theories and research traditions. To see why this is so, we need only trace out certain corollaries of our earlier discussion of scientific rationality. It was observed there that rationality consisted in accepting those research traditions which had the highest problem-solving effectiveness. Now, an approximate determination of the effectiveness of a research tradition can be made *within* the research tradition itself, without reference to any other research tradition. We simply ask whether a research tradition has solved the problems which it set for itself; we ask whether, in the process, it generated any empirical anomalies or conceptual problems. We ask whether, in the course of time, it has managed to expand its domain of explained problems and to minimize the number and importance of its remaining conceptual problems and anomalies.[15] In this way, we can come up with a characterization of the progressiveness (or regressiveness) of the research tradition.

If we did this for all the major research traditions in science, then we should be able to construct something like a progressive ranking of all research traditions at a given time. It is thus possible, at least in principle and perhaps eventually in practice, to be able to compare the progressiveness of different research traditions, *even if those research traditions are utterly incommensurable in terms of the substantive claims they make about the world!*[16]

Hence, even if we could not in principle ever find a way of translating Newtonian mechanics into relativistic mechanics; even if we could never find a way of comparing the substantive claims of twentieth-century particle physics with nineteenth-century atomism; even if, more generally, we could never say that two theories dealt with some of the same problems; it would still be possible in principle for us to make an assessment, *on rational grounds,* of the relative merits of these (or any other) research traditions. This point can be readily generalized by noting that there are many criteria for the comparison of competing theories which do not require any degree of commensurability at the observational level. We might, for instance, compare theories with respect to their internal consistency or their coherence. Equally, we might ask of two (or more) theories, which is the simpler? or which has been refuted? or which yields the more precise predictions? Because such properties (including progressiveness) can be definitely specified, we can say that *the possible incommensurability of theories and research traditions* (so far as their substantive claims about the world are concerned) *does not preclude the existence of comparative appraisals of their acceptability.*[17]

Non-Cumulative Progress

Ever since the appearance of Condorcet's *Sketch Towards a History of the Progress of the Human Mind*, many philosophers and historians of science have developed, at least in outline form, theories of cognitive progress. From Whewell, Peirce, and Duhem through Collingwood, Popper, Reichenbach, Lakatos, Stegmüller, and Kuhn, the search for adequate models of cognitive progress has been, if not commonplace, at least not rare. For all their differences, these models of progress—save Kuhn's[18]—share one common feature: a conviction that *it is only possible to speak of progress if knowledge is acquired through purely cumulative theories.* By "purely cumulative theories," I mean those theories which may *add* to the store of solved problems, but which never fail to solve *all* the problems succesfully solved by their predecessors. Put slightly differently, these thinkers argue that a necessary condition for one theory, T_2, to represent progress over another, T_1, is that T_2 must solve *all* the solved problems of T_1. Although this cumulative conception of progress is usually associated with Popper and Lakatos, it was probably most succinctly formulated by Collingwood when he wrote:

> If thought in its first phase, after solving the initial problems of that phase, is then through solving these, brought up against others which defeat it; and if the second solves these further problems *without losing its hold on the solution of the first,* so that *there is gain without any corresponding loss,* then there is progress. And there can be progress on no other terms. *If there is any loss, the problem of setting loss against gain is insoluble.*[19]

What kind of insolubility is being asserted here? Collingwood never tells us, but presumably what lies behind his concern is the belief that unless the solved problems of one theory form a proper subset of the solved problems of a rival, then we have no way of knowing which theory is the more progressive, for we cannot then reduce progress to a simple additive relation.

Similar concerns motivate the approach of Popper and Lakatos to the nature of progress. In his "requirements for the growth of knowledge," for instance, Popper insists that in order for us to be able to show that a theory is progressive with respect to a competitor, we must be able to show that it entails every fact entailed by its competitor.[20] In the absence of such an entailment, progress (in the Popperian sense) is impossible. Lakatos, for all his quarrels with Popper, takes the same view on this issue: a precondition for saying that a series of theories (i.e., a "research programme") is "progressive" is that each later member in the series must entail all the corroborated content of its predecessor.[21]

Heinz Post has also recently defended the view that new theories always absorb the problem-solving successes of their predecessors. Post goes so far as to make the "claim that, as a matter of empirical historical fact . . . [past] theories always explained the *whole* of [the well-confirmed part of their predecessors] . . . contrary to Kuhn, there is never any loss of successful explanatory power."[22]

The appeal of approaches such as those just mentioned lies in their enormous simplicity. If progress occurred in the manner they require, then we would not have to worry about the counting or the *weighting* of problems. If *all* the previously solved problems in any science are always solved by its latest theories, and if those later theories solve still other problems (regardless of their number or weight), then it is obviously the case that the later theories exhibit progress over the earlier ones. What vitiates this approach to the problem of progress is that the conditions it requires for progress are rarely satisfied in the history of science. As Kuhn, Feyerabend, and others have claimed, there are usually problem losses as well as problem gains associated with the replacement of any older theory by a newer one.[23]

We can get a sense of just how substantial these losses can be by considering a particularly vivid historical example, namely, the shift in geological problems in the early nineteenth century. Prior to Hutton, Cuvier and Lyell, geological theorists had been concerned with a very wide range of empirical problems, among them: how deposits get consolidated into rocks; how the earth originated from celestial matter and slowly acquired its present form; when and where the various animals and plants originated; how the earth retains its heat; the subterraneous origins of volcanos and hot springs; the origin and constitution of igneous rocks; how and when various mineral veins were formed. Solutions, of varying degrees of adequacy, had been offered in the eighteenth century to each of these problems. Yet after 1830, particularly with the emergence of stratigraphy, there were *no* serious geological theories which addressed themselves to *many* of the problems mentioned above. Does that mean (as Popper, Lakatos, Collingwood, and others would have it) that geology was not progressive between 1830 and about 1900 (when many of these issues began to re-emerge)? To jump to that conclusion would be rash, for it would ignore the fact that geological theories after Cuvier and Lyell successfully addressed themselves to a very different range of empirical problems, including those of bio-geography, stratigraphy, climate, erosion, and land-sea distribution. A full analysis of this shift, for which there is not space here, would show that the precision and range of empirical problems that could be solved by mid-nineteenth-century geology (as well of the acuteness of the conceptual and anomalous problems generated) compared favorably to the overall problem-solving success which late eighteenth-century geological theories could claim for themselves. Although this example illustrates more than most just how many problems may fall away from the concerns of a scientific community, the phenomenon it exemplifies is commonplace.

That phenomenon is illustrated within physics by the failure of Newton's optics to solve the problem of refraction in Iceland spar (which had been explained by Huygens' optics), and by the failure of early nineteenth-century caloric theories of heat to explain phenomena of heat convection and generation, problems which had been solved by Count Rumford in the 1790s. Within chemistry, many problems which had been solved by the early theories of elective affinity were not solved by Dalton's later atomistic chemistry.[24] A still better example is afforded by Franklinic electrical theory. Prior to Franklin, one of the central solved problems for electricity was the mutual repulsion of negatively charged electrical bodies. Various theories, especially vorticular ones, had solved this problem by the 1740s. Franklin's own theory, which was widely accepted from the middle to the end of the eighteenth century, never adequately came to grips with this problem.[25]

As these examples show, empirical problems often are either abandoned or relegated to insignificance and any adequate theory of scientific development must presumably allow that such constrictions of the problem domain can, under certain circumstances, be progressive.

I have suggested that such situations can be handled by allowing for the relative importance of various empirical problems. *Knowledge of the relative weight or the relative number of problems can allow us to specify those circumstances under which the growth of knowledge can be progressive even when we lose the capacity to solve certain problems.* In this way, we can avoid the alleged Collingwoodian "insolubility" of how to make a progressive choice between systems, neither of whose problem sets falls neatly inside the other's.[26]

In Defense of "Immature" Science

Both Kuhn and Lakatos are committed to the view that there are two radically different types of science, corresponding roughly to the "early" and "advanced" stages of scientific activity. Although called by different names (for Lakatos, "immature" and "mature" science; for Kuhn, "pre-" and "post-paradigm" science),[27] and defined differently, both writers are committed to the view that the various sciences, at various times, undergo the transition from infancy to majority and that when they do, the rules of the scientific game change substantially. For Kuhn, the transition occurs when one paradigm establishes monopoly over the field and when "normal science" first ensues. For Lakatos, a science reaches maturity when scientists in that field *consistently ignore* both *anomalous problems* and *outside intellectual and social influences,* and focus almost entirely on the mathematical articulation of research programmes. Thus, what chiefly characterizes a mature science, for both Kuhn and Lakatos, is the emergence of paradigms (or research programmes) which are autonomous, and thereby independent from outside criticism. This transition is more than a nominal one; Kuhn and Lakatos alike insist that mature science is more progressive, more genuinely scientific than its immature counterpart.

There are several troubling aspects to the concept of a mature science (at least as developed by Kuhn and Lakatos). The suggestion that every (or even that any) science undergoes a *permanent* transition of the kind which Kuhn and Lakatos describe does not square with what we know about the evolution of science. Kuhn can point to no major science in which paradigm monopoly has been the rule, nor in which foundational debate has been absent. Lakatos, for his part, has identified no (physical) science in which the disdain for anomaly and the indifference to extra-programmatic conceptual problems have been the prevailing features. As a result, it is extremely unclear whether the notion of a "mature" science finds any exemplification whatsoever in the history of science.

Even if mature sciences existed, the Kuhn-Lakatos thesis that they would be intrinsically more progressive and more scientific than "immature" ones has not been established. Kuhn has not shown that more empirical problems necessarily would be solved if one paradigm tyrannizes a scientific domain. Lakatos has not made a convincing case for his claim that autonomous, anomaly ignoring research programmes are likely to be more progressive than non-autonomous, anomaly-recognizing ones.[28] In the absence of cogent arguments for the greater rationality of mature science, we can only conclude that the preference voiced by Kuhn and Lakatos for mature sciences is without foundation.

A third difficulty with the doctrine of mature and immature sciences is the scope it allows the builder of any model of scientific rationality to dismiss as irrelevant any historical counter-examples to his model. Since the models are chiefly designed as replicas of "mature science," any actual scientific examples which fail to fit the models can be explained away as proto- or pseudo-science rather than being regarded as genuine exceptions to the models. The mature-immature dichotomy is thus *methodologically suspect* because it effectively renders these models of scientific rationality immune from empirical criticism.[29]

In arguing against the existence and the desirability of mature science, so construed, I am manifestly not claiming that the later stages of a science exhibit all the structural and methodological features of theories in earlier stages. We may yet find a characterization of mature science which will do justice both to history and to rationality.[30] But the concept of mature science as conceived by Lakatos and Kuhn unfortunately does neither.

Part Two

Applications

Chapter Five
History and Philosophy of Science

> ***Philosophy of science without history of science is empty; history of science without philosophy of science is blind.*** I. LAKATOS (1971), p. 91

Because the stimulus for developing the model in Part One came chiefly from the writings of historians and philosophers of science, it is appropriate to begin exploring the ramifications of that model by examining its consequences for this field (or fields). To say as much is already to indicate what must be one of our core concerns; for the parenthetical expression above stresses just how unclear scholars still are about whether history of science and philosophy of science are two distinct domains or whether, as some writers allege, they are so intimately connected as to be one field, incapable of meaningful separation. Couched in these terms, the issue might seem to be largely verbal—one of those boring disputes about where the boundaries of one discipline end and another begins. But in this case, there are some substantive issues that hang on the separability of history and philosophy of science. Questions about the aims, the methods of inquiry and the manner of legitimating both historical and philosophical claims are bound up in the question: are history of science and philosophy of science autonomous enterprises?

The standard view, of course, regards history of science and philosophy of science as radically different, if perhaps sometimes complementary, ways of studying science. The historian, on this view, is dealing with facts and data, seeking to arrange them into a convincing and coherent tale about how scientific ideas have evolved. Philosophy of science, by contrast, is commonly perceived as a normative, evaluational and largely *a priori* investigation of how science ought to proceed. On this view, the gap between history and philosophy of science is as broad as, indeed is illustrative of, the divide between matters of fact and matters of value. History is irrelevant to the philosopher because he is not concerned about what science has been, but rather how it should be. Philosophy is irrelevant to the historian because it is not his job to make normative judgments about the figures he studies.

Research in the last twenty years has done much to highlight the weaknesses in the standard account. Agassi,[1] Grünbaum[2] and others have shown how much writing in the history of science is laden with implicit philosophical assumptions, assumptions which decisively determine the character of the history that is produced. (To take an overly simple example, if a historian is convinced that experiments can be the only decisive grounds for abandoning a theory, then his history will tend to focus exclusively on so-called crucial experiments.) The thesis is not merely that philosophical assumptions *have* influenced historical scholarship, but that they *must* do so, because history (like science) has no neutral data, and because the treatment of any particular historical episode is going to be influenced to some degree by one's prior philosophical conceptions of what is important in science.

The correlative case for the philosophy of science has been argued with equal fervor by a number of thinkers, including Whewell, Hanson, Kuhn, Toulmin, Lakatos, McMullin, and Feyerabend.[3] While granting that the aim of philosophical inqury is the production of a set of norms (e.g., for choosing between competing theories), these critics of the standard view point out that any philosophical theory of science which failed completely to square with the history of science would be deemed unacceptable. Confronted with a philosopher's account of, say, rational theory acceptance which entailed that the whole of the history of science was irrational, we would tend to view that as a *reductio ad absurdum* of the theory of rationality rather than as a demonstration that science itself had been a sequence of entirely irrational preferences.

If these critics are right, there are relations of mutual dependence between history of science and philosophy of science which make nonsense of any attempt to allow them an autonomous development. But there are, *prima facie,* some difficulties with an integrated view of history and philosophy of science, difficulties so acute that most thinkers have remained unconvinced by the claim of mutual dependence. Foremost among these difficulties is the *vicious circularity* which it seemingly entails. If the writing of history of science presupposes a philosophy of science and if a philosophy of science is then to be authenticated by its capacity to lay bare the rationality held to be implicit in the history of science, how can we avoid automatic self-authentication, since the history we write will presuppose the very philosophy which the written history will allegedly test? Other difficulties abound. If, as seems likely, virtually all the available philosophies of science can do scant justice to the history of science, why should the historian of science take them seriously as theoretical tools for organizing his research? Similarly, if most history of science has been written utilizing discredited philosophical models of science, why should the philosopher feel constrained to test his carefully worked out models against historical "data" which were collected under the aegis of a naive or opposing philosophy of science? There are some slightly more technical troubles as well. Even if we grant that in some sense the actual course of science should have some bearing on the philosophy of science, how close does the fit between actual history and its normative reconstruction have to be? Because no one, neither historian nor philosopher, is committed to the view that the *whole* of science is rational,[4] why should the philosopher be disturbed if, on his account, many episodes in the history of scientific ideas turn out to have irrational elements?

These are large, and still unanswered questions. The aim of this chapter is to provide some of those answers.

The Role of History in the Philosophy of Science

There are, of course, already certain areas of philosophy of science where a significant empirical input from the sciences is taken for granted. To take but two examples, the philosophy of space and time and the philosophy of biology are universally acknowledged to draw heavily upon the recent state of natural sciences. But in that portion of philosophy of science concerned with general methodology (e.g., with norms for theory appraisal and assessment), there is still a widespread discomfiture with the suggestion that empirical data about the evolution of science are relevant or decisive.

Before we attempt to resolve these questions, it will be helpful to remind ourselves of one elementary but crucial distinction which is germane to this discussion: specifically, the distinction between the history of science itself (which, at a first approximation, can be regarded as the chronologically ordered class of beliefs of former scientists) and writing *about* the history of science (i.e., the descriptive and explanatory statements which historians make about science). Vital though the distinction is, it is often forgotten—in part, presumably, because speakers of the English language use the same name for both. Because some of the confusion about the relations between history of science and philosophy of science derives from an equivocation on these two different senses, I shall use "HOS_1" to refer to the actual past of science and "HOS_2" to refer to the writings of historians about that past.

A fresh version of the traditional case for the autonomy of philosophy of science (in the sense of general methodology) from HOS_1 was recently published by Ronald Giere.[5] His approach involves the familiar insistence that philosophy of science is normative, and because one cannot derive norms from "facts," he does not see how the history of science could be relevant to philosophy. He goes on to say that although a philosopher might come to some new insights by studying HOS_1, such study is no part of the authentication or the validation of those insights, since (Giere tells us) they could have been discovered in any case without those historical examples. Finally, Giere insists that the philosopher must not become a slave to HOS_1 because one of his primary roles is to criticize the theories of the past. For such criticism to have any bite, we must have independent, nonhistorical grounds for it.

Giere's views (which are, as he says, "fairly representative of the majority of philosophers of science"[6]) seem plausible at first glance. But they quickly begin to crumble under detailed examination. As he himself concedes, if any philosophy of science were to entail that virtually all our previous scientific judgments were irrational, then we would have grave doubts about the "claim [of that philosophy of science] to be talking about scientific theories."[7] Precisely because "philosophical theses cannot be completely *a priori,*" they must capture some of our prephilosophical hunches about which theories are rational and which are not.[8] If these hunches do not come from HOS_1, where do we obtain them? Giere's answer gives the game away: it is, says Giere, to recent and contemporary science that the philosopher of science must look for inspiration and legitimation. Giere fails to see that his use of current "actual scientific practice" (his examples are quantum mechanics, molecular biology, and contemporary psychology[9]) is itself the invocation of HOS_1 to adjudicate philosophical claims. The fact that a scientific theory is still believed and is currently undergoing development scarcely makes that theory ahistorical. Every example which a Gierean philosopher of science will discuss is drawn from the past, from history. Giere's own historical preferences may be for the recent past, but they are nonetheless historical for all that.

What lies behind Giere's point, I believe, is a recognition that much of HOS_2 (note the subscript) focusses on the distant past, and that there are as yet all too few historical accounts of recent HOS_1. But the fact that philosophy of science can dispense with HOS_2 does not militate against the parasitic dependence of philosophy of science upon HOS_1. What is clear, therefore, is that a resolution of the normative/descriptive paradox is as crucial for those philosophies which are grounded in contemporary science as it is for those philosophies which look back beyond our own time. Needless to say, a *tu quoque* argument of this type does not solve that core problem; to the contrary, it accentuates its importance by revealing its universality.

I shall propose one possible way out of the paradox. Let us begin by returning to the distinction between HOS_1 and HOS_2. Within HOS_1, there is, I shall claim, a subclass of cases of theory-acceptance and theory-rejection about which most scientifically educated persons have strong (and similar) normative intuitions. This class would probably include within it many (perhaps even all) of the following: (1) it was rational to accept Newtonian mechanics and to reject Aristotelian mechanics by, say, 1800; (2) it was rational for physicians to reject homeopathy and to accept the tradition of pharmacological medicine by, say, 1900; (3) it was rational by 1890 to reject the view that heat was a fluid; (4) it was irrational after 1920 to believe that the chemical atom had no parts; (5) it was irrational to believe after 1750 that light moved infinitely fast; (6); it was rational to accept the general theory of relativity after 1925; (7) it was irrational after 1830 to accept the biblical chronology as a literal account of earth history.

The precise dates here are not important, nor yet is any single item on the list. What I shall maintain, however, is that there is a widely held set of normative judgments similar to the ones above. This set, constitutes what I shall call *our preferred pre-analytic intuitions about scientific rationality* (or "PI," for short). (This set is a very small subset of all our beliefs about HOS_1.) Our convictions about the rationality or irrationality of such episodes are clearer and more firmly rooted than any of our overt and explicit theories about rationality in the abstract. Particularly decisive here are those theories and research traditions which have been the most global and the most influential, i.e., those which have for long epochs provided the motivation and presuppositions for a wide range of detailed theorizing. Any model of rationality which led to the conclusion that the acceptance of most of these doctrines were ill-founded would have few claims on our loyalty.[10] As a result, our intuitions about such cases can function as decisive *touchstones* for appraising and evaluating different normative models of rationality, since we may say that it is a necessary condition of any acceptable model of rationality that it square with (at least some of) our PIs.

How, in practice, can such episodes test a putative model of rationality? In outline, the procedure is a simple one. Any philosophical model will specify certain parameters as being relevant to the acceptance of a theory (e.g., in the case of the model in Part One, these would be the solved, anomalous and conceptual problems exhibited by any theory and its competitors). Historical research into the case in hand would indicate what their values should be. Once these values are specified, the model should lead us to a determination of the historical rationality of accepting the theory in question. If the evaluation issuing from the model accords with our pre-analytic intuitions, then the latter provide support for the model; if, on the other hand, the model's verdict contradicts our pre-analytic judgments, then the model is in serious jeopardy.

In the extreme case, a proposed model of rationality would be justifiably dismissed out of hand if, when applied to the cases involved in PI, it entailed that all our intuitions were incorrect, for it would have failed to capture the very rationality it was designed to explicate. We should be very explicit about what we are committing ourselves to in taking this approach: (1) *that at least certain specified developments in the history of science were rational;* and (2) *that the test of any putative model of rational choice is whether it can explicate the rationality assumed to be inherent in these developments.* Claim (1), modest though it is, remains entirely a matter of faith since there is, in principle, no way we could prove these cases were rational, for our criterion of rationality itself will take their rationality for granted.

We have thus far mentioned only the *extremal* case in which a methodology is discredited by every element of PI: though an extreme, it is common enough (indeed many contemporary philosophies of science are supported by *none* of the cases above). Even so, we can move beyond the extremal case to claim more generally that *the degree of adequacy of any theory of scientific appraisal is proportional to how many of the PIs it can do justice to.* The more of our deep intuitions a model of rationality can reconstruct, the more confident will we be that it is a sound explication of what we mean by "rationality."

As natural as the proposal to utilize history of science as a testing ground for philosophical models of rational choice might seem, there are probably those purists who regard it as unseemly that philosophy should have to look beyond itself and its own argumentative strategies for legitimation. But where, *within* philosophy, can one find the appropriate decision criteria? Suppose that we are confronted with two competing models of rationality, MR_1 and MR_2 (each of which is internally consistent). How, in principle, could we make a rational, *philosophical* choice between them? Since both MR_1 and MR_2 purport to define the conditions for rational choice, any choice between them would presuppose the validity of one or the other model (or perhaps yet a third model). We clearly have a serious meta-level problem which can only be solved by testing the competing models against something besides a theory of rational choice itself. The proposal in this chapter is that our criterion for choice between competing theories of rationality should involve evaluating such models against those archetypal cases of rationality (PI) which we find in HOS_1.

This proposal for the authentication of philosophical claims about scientific rationality makes it clear that philosophy of science depends in two crucial respects upon the history of science. In the first respect, it aims to explicate the criteria of rationality implicit in our preferred intuitions about certain cases within HOS_1. In the second respect, the authentication of any philosophical model requires careful research in HOS_2 in order to assess the applicability of that model to the cases which constitute PI.

But does this approach then make the philosophy of science merely descriptive and rob it of any critical force? The general answer is no. About *most* episodes in HOS_1, we have no strong, widely shared, pre-analytic convictions. Indeed, the chief point of constructing a model of rationality is to use it to get clarification about the "fuzzy" cases (which are the overwhelming majority). With respect to the latter, the philosopher's judgment—based on a model of rationality authenticated by the set of PIs—must take precedence over whatever weakly held, pre-analytic hunches we may have about them. As in ethics, so in philosophy of science: we invoke an elaborate set of norms, not to explain the obvious cases of normative evaluation (we do not need formal ethics to tell us whether the murder of a healthy child is moral), but rather to aid us in that larger set of cases where our pre-analytic judgments are unclear.

Hence, the philosophy of science is both descriptive and normative, both empirical and *a priori,* but with respect to different types of historical cases.

There are doubtless other ways in which the HOS_1 might be helpful to the philosopher of science, ranging from providing illustrations of philosophical claims to offering heuristic guidance for the handling of specific issues.[11] But the philosopher does not need HOS_1 for these ends. The one and only point where he cannot dispense with HOS_1 is when it comes to deciding whether his would-be theories of rationality are, in fact, theories of rationality.

Imre Lakatos has already made a suggestion akin to mine about utilizing HOS_1 to "test" any model of scientific rationality. There are, however, important differences of substance between our approaches which are worth exploring. Essentially, Lakatos' proposal is that the best model of scientific rationality is the one which, when applied to HOS_1, will allow us to represent *the largest portion* of scientific history as a rational enterprise. It is, in short, not a small set of cases about which we have strong intuitions (as I propose), but the whole of the history of science (i.e., HOS_1) which becomes the criterion for choosing between different models of rationality.[12] Lakatos' approach strikes me as counter-intuitive for a very simple reason: if we take his proposal seriously, then *the best possible model of rationality would be that one which resulted in the judgment that every decision ever made in the history of science was rational.*[13] This seems a curious ideal to strive after, for just as we are convinced that some scientific choices have been rational ones, we are equally convinced (given "human nature") that not all of them have been rational. Any model of rationality which made the *whole* of science rational would be as suspect as those models which make *none* of science rational. My suggestion about the use of the set PI as a device for testing the models of rationality is an effort to find a plausible middle ground between such extremes.

The Role of Norms in the History of Science

If the focus in the previous section was on the relations of philosophy and HOS_1, the concern here is with the connections, if any, between HOS_2 and the philosophy of science.[14] This case is a more complex one, for the points at which valuational elements enter HOS_2 are more subtle and more implicit than in the reverse case. We will examine two very different points of contact: in constructing a historical narrative and in offering historical explanations.

Norms in Historical Narration

As Agassi pointed out in his classic study of the historiography of science,[15] every working historian of science must, in sifting and arranging his data, make many assumptions about the character of science. He must assume, among other things, that there were scientists, and he must be able to distinguish those of their activities which were scientific (and therefore relevant for inclusion in his narrative) from those which were not. Even among the class of scientific activities, the historian must prune and select, since there are acute, practical limitations on the completeness which HOS_2 can achieve. He must decide, for instance, how much importance to give to discussing a scientist's experiments, his theories, his laboratory journals, his lecture notes, the books in his library and the like. In principle, the historian could presumably make these decisions by some randomizing device; but, in practice, *what guides the historian's choices are a set of assumptions about what is most important to the doing of science.* Philosophical and normative elements inevitably enter at this stage. How much importance a historian assigns to discussions of experiment will depend on how important he believes experiments to be to scientific development. The significance he assigns to the religious or metaphysical background of a scientist will depend, again, upon the historian's conviction about how decisive those elements are in scientific deliberation.

Not surprisingly, historians with different "images" of science will give radically divergent accounts of the same episodes (a phenomenon probably exhibited most vividly in Galileo scholarship—where we find Marxist, idealist, empiricist, instrumentalist, and pragmatic accounts of the "same" scientific achievement). There is nothing wrong in this; or perhaps we should say that, wrong or not, it is inevitable that any historian's account of science is going to be colored by his views about how science works. Such "coloring" only becomes invidious when the motivating philosophy of science is implicit and uncritically utilized, or when its existence is denied by the historian who imagines that he is free from any normative biases.

But we can go further than this. It is the historian's intellectual—even moral—obligation not only to be self-conscious about the kinds of norms he is applying, but also to see to it that *he is utilizing the best available set of norms.* How can he make that choice? By accepting that model of rationality (or perhaps those models if we can find more than one satisfying the appropriate conditions) which does the greatest justice to our PIs about HOS_1. It is with this step that we complete the circle connecting the history and the philosophy of science. *The task of the historian of science,* so conceived, *is to write an account (HOS_2) of episodes in the history of science (HOS_1) utilizing as his criteria of narrative selection and weighting those norms contained in that philosophical model which is most nearly adequate to representing PI.* To do any less than this, to utilize a half-conscious or less than adequate model of science for writing the history of science, is as intellectually irresponsible as deliberately ignoring the evidence.

Many historians will doubtless agree that this is the ideal; if it is rarely achieved that is primarily because the models offered up by philosophers seem to be even less adequate than the historian's own half-articulated views about the norms of scientific evaluation. But, inductive evidence to the contrary notwithstanding, the historian should not assume that every philosophical model of rationality is incapable of illuminating history.

Norms in Historical Explanation

Thus far, we have spoken only of the way in which philosophical beliefs about science influence the historian's decisions about what factors to include in his narrative accounts. But there is a second, deeper level at which philosophical or normative judgments ineluctably enter into HOS_2—at the level of historical understanding and historical explanations. Though by no means the only goal of HOS_2, one of its primary functions is to explain why various experiments, theories, and research traditions were accepted, rejected or modified in the ways that they were. Any serious study in the history of scientific ideas will be replete with explanations of such factors. Normative evaluations are crucially involved in all such explanations—not as explicit premises, but as their ground. Consider a very typical example:

Q_1: Why did Newton reject Descartes' vortex theory of planetary motion?

A_1: Because Newton correctly judged that Descartes' theory was grossly *incompatible* with data about the velocities and positions of the planets.

Clearly, the answer is meant to explain Newton's rejection of the vorticular hypothesis. But suppose we go a step further and ask:

Q_2: Why should Newton reject a theory that is grossly incompatible with the data?

The question itself seems peculiar; it does so because historians take it for granted that it was reasonable in Newton's time to insist that theories be compatible with the data, and that if one can show that someone's action was reasonable (under the circumstances), then there is nothing left to explain—our explanatory task is finished. Questions like Q_2 seem superfluous. The history of science (HOS_2) abounds with such cases: the historian explains why a scientist accepted a certain idea by showing that the scientist deduced it from a prior belief; he explains why a scientist performed an experiment by showing that it would test a theory the scientist was considering.

In *all* such cases, we are implicitly relying on a conception of, "what it would be reasonable to do in the circumstances." To see that this is what is involved, consider a perverse "explanation" along the following lines:

Q_3: Why did Jones accept the evolutionary hypothesis?

A_3: Because all the evidence was *against* it.

Clearly, something is wrong here. In fact, the answer might be true. If, for instance, we also know that Jones was a resolute *iconoclast* who always denied the evidence of his senses, then the explanation would become convincing (although, of course, we might still want to know what caused Jones' iconoclasm). But as it stands, A_3 carries no explanatory force. It fails to do so because the reason it offers for accepting evolutionary theory seems to be no legitimate reason at all. If, on the other hand, our answer had been:

A_3^1: Because all the evidence *supported* it, we would be reasonably content with the answer (provided, of course, there was historical evidence for A_3^1).

The point is that the historian's explanations continually invoke canons of rationality and plausibility, and thereby presuppose a huge amount of normative machinery. And here, as with norms of selection, the historian should see to it that the norms of rationality he invokes are the best ones available.

Other vital dimensions of historical research similarly require the use of norms about rational belief and rational action. To an extent rarely appreciated by nonhistorians (who often imagine that the historian is a mere reporter of events), studying the history of ideas—scientific or otherwise—involves a great deal of creative imagination. Scientists rarely leave full accounts of how they came to make their discoveries; even when they do, such accounts are often unreliable, because constructed long after the fact. The task confronting the historian is often that of conjecturally recreating lines of argument and influence which lay behind the conclusions which a scientist explicitly propounds. This task of reconstrution is utterly impossible unless the historian has a very subtle sense of what kinds of arguments would be plausible in a given situation. Thus here, as with narration and explanation, the historian's task requires that he possesses a theory (implicitly or explicitly) about rational belief and rational action.

Rational Appraisal and "Rational Reconstruction"

What has prevented many historians from seeing the force of these arguments is a fear that subscription by them to any *contemporary* model of rationality will lead to the anachronistic importation into the past of criteria of rational choice which are not relevant to the historical circumstances.[16] Precisely because he knows that the norms of rational evaluation change through time, the historian worries about the appropriateness of transposing our contemporary philosophical insights—assuming that sound ones could be found—into an epoch and a culture to which they are foreign. He has a right to insist that any theory of norms, if it is to be applied historically, must take into account the fact that previous scientists had norms of their own (often different from ours) which cannot be ignored in explaining their cognitive stances with respect to the theories of the day. Because no philosophical model of rationality has made any concessions to the norms of the past, the historian has been understandably loathe to utilize such models.

Indeed, perhaps the chief stumbling block to the historian's admission of the relevance of philosophy to HOS_2 has been the flagrant disregard for HOS_1 exhibited by many of those very philosophers (especially Lakatos, Feyerabend, and Agassi) who have argued most loudly for the dependence of HOS_2 upon philosophy.[17] This disregard extends, not merely to their misuse of historical data, but is deeply grounded in their convictions about the aims of a philosophically based history of science, convictions which sometimes subordinate historical veracity to the desire to score philosophical points.

These issues probably emerge most clearly in Lakatos' "theory of rational reconstruction," itself a theory about the role of philosophy of science in writing HOS_2.[18] Lakatos purports "to explain *how* the historiography of science should learn from the philosophy of science."[19] The rational reconstructions of the past, which Lakatos urges the philosopher to undertake, bear a very curious and ambiguous relation to the actual episodes of which they are the purported reconstruction.

As Lakatos insists, the process of preparing the "internal" history or rational reconstruction of a historical episode is not really an empirical task at all. One *"invents"* or "radically improves" on the actual historical record in order to "rationally reconstruct" it.[20] In this rational reconstruction, one tells history *as it ought* to have happened. The actual beliefs of the historical agents whose names figure in the story are ignored or often deliberately distorted. Lakatos here is *not* making the point that the historian is inevitably selective in the data which he mentions. He is rather making the very different claim that the "rational historian" should construct *a priori* an account of how a particular episode should have occurred. There need be *no* resemblance whatever between the "internal" account so constructed and the actual exigencies of the case under examination.[21]

If this sounds extreme, one of Lakatos' examples makes clear just how far he is prepared to move away from the historical record. When discussing Bohr's theory of the electron, for instance, Lakatos points out that Bohr had not, by 1913, even conceived the idea of electron spin. "Nonetheless," insists Lakatos, "the historian, describing with hindsight the Bohrian programme, should include electron spin in it, since electron spin fits naturally in the original outline of the programme. Bohr might have referred to it in 1913."[22] On this criterion, anything whatever that a historical figure *might* have said (i.e., presumably anything that is consistent with his "research programme") can attributed by the historian to that figure. The honest Lakatosian historian must, of course, "indicate *in the footnotes* how actual history 'misbehaved',"[23] but the reconstruction itself is by no means limited to the actual beliefs of historical agents. Indeed, the liberties which the rational reconstructionist is permitted go well beyond filling in beliefs which are consistent with a thinker's research programme. He may often, too, ignore or even repudiate the standards of rationality of a historical figure if he finds them uncongenial. In discussing the work of the chemist Prout, for example, Lakatos urges the historian to ignore one of Prout's basic beliefs about the experimental well-foundedness of his hypothesis about elemental composition.[24]

Once an episode has been so re-cast by the rational reconstructionist, he proceeds to appraise its rationality, according to an appropriate model of rational choice. Whatever the outcome of that appraisal, however, *the historical episode itself remains untouched and unexplained*—except to the extent of its faithfulness to the *a priori* reconstruction (an isomorphism that will, in the nature of the case, scarcely ever exist except in limited fashion).[25]

Lakatos defends this theory of rational reconstruction by arguing that *"history without some bias is impossible."*[26] There is surely a difference, however, between having a theoretical bias (i.e., selecting and interpreting historical events "in a normative way"[27]) and consciously and deliberately falsifying the historical record. Lakatos nowhere establishes the necessity (or the desirability) of making a reconstruction of the past which involves an intentional warping of the historical record. Indeed, the fact that Lakatos assumes the possibility of comparing a "reconstruction" of an episode with its "actual history"[28] shows that Lakatos himself believes that history does not have to be "fabricated" to be understood.

I want to dissociate my own model of scientific rationality as vigorously as possible from those of Lakatos and the other rational reconstructionists. Like them, I believe that the appraisal of the rationality of historical episodes is an essential task for the historian of scientific ideas. But there the similarity ends.[29] Unlike the rational reconstructionist, I insist that it must be actual episodes, not some figment of our imagination, whose rationality we assess. Unlike them, I argue that the actual beliefs of historical agents, *and* the canons of rational belief of their epoch, must be scrupulously attended to. In contrast to the reconstructionists, I object to the invention of historical figures and the fabrication of historical beliefs in order to score philosophical points or to teach philosophical lessons.[30] If the *philosopher* would learn something from history, he must make himself a servant to it—at least to the extent of dealing with actual cases and actual beliefs. And if the *historian* is to find any philosophical model relevant for his own work, that model must allow for the evolving character of rationality itself. I have already claimed that the model developed in Part One can succeed in doing just that.

Chapter Six
The History of Ideas

Though the gap seems small, there is no chasm that more needs bridging than that between the historian of ideas and the historian of science. T. S. KUHN (1968), p. 78

The work of too many professional historians is diminished by an anti-rational obsession—by an intense prejudice against method, logic and science. D. FISCHER (1970), p. xxi

The history of ideas or, as it is often called, intellectual history, is among the oldest genres of historical writing. The presumptions which motivate it, namely, that what our ancestors *thought* is as interesting as what they *did*, that their ideas were as important as their wars and rulers, have their roots deep in antiquity; indeed, many of the earliest extant historical writings are concerned with what we would now call the history of ideas. In recent times, particularly in the nineteenth century, studies of the history of thought, cultural history, the evolution of ideas and doctrines formed a large share of the historical literature. In our own time, by contrast, the history of ideas is regarded in many quarters as passé and irrelevant, as a discipline with outmoded presuppositions and outrageous ambitions. Many historians see intellectual history as an anachronistic excrescence on the scholarly and ideological integrity of their field. Because the bulk of this chapter (and in certain respects the entire essay) is an effort to stress the importance of the history of ideas—at least of a certain type of history of ideas—it is probably wise to begin by surveying some of the reasons for its present disrepute.

There are several, frequently cited complaints directed against intellectual history:

1. *that it is "elitist":* not because most people do not think, but rather because we only have historical records about the "thoughts" of a tiny fraction of the persons in any society (namely, those who were both literate and fairly prolific).

2. *that it assumes ideas have an independent reality:* it is, so the critics stress, *"people* who have ideas." People live in societies with certain economic, political and social characteristics which condition or even cause their ideas. Intellectual history, to the extent that it abstracts ideas from their broader social surroundings, distorts the historical record.

3. *that ideas are a far less potent source of change than the underlying socio-economic "realities":* on this view ideas (in form of "ideologies") merely mirror the material condition of the society and serve only as tokens for the class conflict between the warring factions. To focus on the evolution of ideas is to misplace the genuine causes of historical change.

4. *that the history of ideas, because it is "impressionistic" and not readily quantifiable, is out of step with the move towards "scientific" history.*

I shall postpone any direct comment on these well-known quibbles with intellectual history. It was important to state them early, however, in order to underscore the differences between these standard criticisms of the history of ideas and those reservations which I shall be voicing. All the above are objections in principle to *any* type of intellectual history; they seek to cast doubts on any effort to study the evolution of ideas (except within a broader socio-economic context). My own reservations, which I shall discuss at length, are qualms about the assumptions currently underlying certain types of intellectual history. In brief, I shall argue that much intellectual history, as currently practiced, is too discipline-oriented in its approach, too insensitive to the historical dynamics of intellectual problems, and more preoccupied with chronology and exegesis than with explanation—which should be its central object. But all these defects are remediable. My claim will be that there are ways of doing the history of ideas which are not only intellectually well-founded, but highly relevant as well. After describing what I believe to be, at least in its essentials, an adequate model for a historiography of ideas, I shall return to points (1) through (4) above in order to see how cogent they are in the face of a more complex conception of intellectual history.

Disciplinary Autonomy and the History of Ideas

Without any doubt, one of the most restrictive features of much intellectual history is its discipline-bound manner of presentation. We have historians of philosophy, historians of science, historians of theology, each generally assuming that the ideas with which they are concerned have no crucial cross-disciplinary dependencies. The tendency toward specialization extends even into single disciples. Philosophers write histories of ethics, histories of epistemology, and histories of logic. Scientists write histories of analytic chemistry, histories of physical optics, even histories of x-ray crystallography. Theologians give us histories of eschatology, histories of natural theology, and histories of eucharistic doctrine. There is nothing surprising in all this. Practitioners of a contemporary speciality have a natural, perhaps inevitable, curiosity about their predecessors. Nor is there *necessarily* anything suspect about the high degree of specialization which we see in much contemporary writing about intellectual history. But in practice, if not in theory, this manifold division of labor between the various disciplines has exerted a deleterious effect on the writing of intellectual history, because the assumption of (relative) disciplinary autonomy has tended to blind many historians of ideas to the single most striking fact about the history of thought, *its integrative character*.

Up to, and even including, our own time, leading intellectuals have been concerned simultaneously with a broad spectrum of problems and issues, ranging from the highly specific and technical to the very general and abstract. As I showed in Part One, rational appraisal has generally been construed by our predecessors as a process of finding maximally adequate solutions to a divergent range of compelling intellectual problems, problems which, moreover, occur in several diverse disciplines.[1] The evolution of ideas, and the problems to which those ideas provide solutions, is necessarily *an interdisciplinary process.* Historians of ideas, scientific or nonscientific, ignore this integrative tendency at their peril.

Yet they do ignore it. The vast majority of contemporary histories of science and histories of philosophy pay no more than lip service to the mutual interpenetration of "scientific" and "philosophical" doctrines and problems. Equally, one is hard pressed to find any history of social or political theory which is fully alive to the high degree of historical interaction between the "soft" and the "hard" sciences.

If the nature of the interaction between the various disciplines were just a kind of "spill-over" effect, whereby ideas from one domain only occasionally penetrated into another, the tendency to write disciplinary histories of ideas would be excusable. But the fact of the matter is (if we extrapolate from the best recent scholarship) that there is—or at least has been—a continuous process of interpenetration and legitimation going on between the intellectual structures of the various disciplines. Thus, the problems of seventeenth- and eighteenth-century metaphysics were posed by the new "mechanistic science" and make no sense except when seen against that background. The problems of nineteenth-century social theory and aesthetics were the by-product of a confluence of scientific, technological, and epistemological developments which provided both the model for, and the legitimation of, a succession of theories about social structure and aesthetic perception.

What has led otherwise subtle and sophisticated scholars to ignore so many of these interconnections? Why, more specifically, has the chasm "between the historian of ideas and the historian of science" (to which Kuhn refers in the passage at the head of this chapter) developed? The core of the answer is provided, ironically, in Kuhn's own work. Although bemoaning the failure of historians to see the connections between scientific and nonscientific ideas, Kuhn himself articulates a now well-known model of scientific development which, in its essentials, denies the existence of any significant degree of interaction. It is Kuhn, for instance, who writes that: "the practitioners of a mature science are effectively insulated from the cultural milieu in which they live their extra-professional lives."[2] It is Kuhn, again, who insists that: "the development of an individual technical speciality can be understood without going beyond the literature of that speciality and a few of its near neighbors."[3]

Such tensions between the historian's aspirations and his convictions are so familiar as to be commonplace.[4] While insisting that we *ought* to look for intellectual connections between disciplines, when it comes to the discipline he knows best, the historian often proceeds to write about its history as if it were almost completely isolated from everything else! He does not seem to realize that so long as we retain a model of strict disciplinary autonomy, then the realization of an interdisciplinary history of ideas will forever elude us.

Ideas and Their Problem Contexts

A related and persistent failing of much scholarship in the history of ideas is the tendency to ignore the *problems* which have motivated the construction of great intellectual systems of the past. Too often, the historian of ideas sees his function primarily as that of setting out the systematic interconnections between the beliefs of a thinker or group of thinkers on a closely related family of issues; this is a subtle task, which involves revealing the threads of reasoning whereby our predecessors came to hold the beliefs they did. But this is to tell only half the story, even when done well. Systems of thought are not merely logical links between propositions. They are that, but they are also attempts to resolve what are perceived as important problems. To write about the history of conceptual systems without ceaselessly identifying the problems which motivated those systems is drastically to misconstrue the nature of cognitive activity.[5] To give, say, a detailed exegesis of Locke's empiricism or Engel's dialectical materialism without carefully identifying the empirical and conceptual problems that those doctrines were designed to resolve is not unlike playing one of those parlour games in which one is given an answer (often a bizarre one), without knowing the question to which it is an answer! One can only understand a system of ideas when one knows, in detail, the problems to which it was addressed.

If it seems difficult to imagine that this commonplace is more often ignored than observed, consider a pair of examples. For several hundred years, historians of ideas have been writing about Cartesian philosophy. Literally hundreds of books and thousands of articles have been written about Cartesian dualism, about the method of doubt in Descartes, about the *cogito* argument, and about Descartes' borrowings from his predecessors. Yet it is only in the last generation that scholars, such as Gilson and Popkin,[6] have begun to shed any useful light on Descartes' problem situation and orientation. Only now can we begin to see why Descartes' philosophy sometimes takes those curious twists and turns that made so little sense when scholars were insensitive to the actual problems with which that philosophy grappled.

A second example is provided by the vast exegetical literature dealing with John Stuart Mill's influential views on epistemology, logic, and political philosophy. As extensive as it is, we still have almost no sense of Mill's problem situation. Why, for instance, did he devote so much energy to reviving the methods of enumerative and eliminative induction? What were the specific problems within the social sciences for which his well-known "historical method" was meant to provide a solution? What was his motivation for classifying the sciences in the manner he did? Much of the most careful scholarship on Mill skirts these (and other similar) questions about the problems Mill was tackling.

Even when intellectual historians recognize that systems of thought have their roots in problems, they sometimes tend to adopt an ossified and unilluminating notion of what a problem is. Showing less sensitivity to historical process and conceptual nuance than one might expect, many scholars write as if problems have an unchanging identity through time, a *perennial* character.[7] How often does one see references in the history of philosophy to *the* problem of substance, *the* problem of induction, *the* mind-body problem, *the* problem of free will, *the* problem of universals? Similarly, historians of science speak of *the* problem of combustion, *the* problem of life, or *the* problem of free fall. In each case, these problems have not remained static through time. Hume's problem of induction was very different from Mill's and both are very different from our version of it.[8] There are times when two thinkers do address the same problem or set of problems; but that must be shown rather than idly presupposed. To *assume* problem identity through time is, for the historian of ideas, the first step on the road to what may be a most serious falsification of the historical record, for when we misconceive the precise character of a thinker's problems, we are apt to misunderstand the nature of the solutions he proposes.

Many historians are quite forthright in their insistence that intellectual problems are unchanging. Leonard Nelson, for instance, goes so far as to claim that it would be *impossible* to write the history of philosophy without assuming the identity of problems through time. On Nelson's analysis, solutions may change, but problems cannot.[9] Nelson's approach borders on the perverse. To imagine—as Nelson would have it—that medieval theology, or seventeenth-century physics, or the recent emergence of the social sciences produced no *new* problems for the philosophical tradition requires a massive repudiation of much of the best scholarship of the last 150 years.

The emphasis I have given to the importance of a problem-oriented approach to intellectual history echoes Collingwood's insistence that the historian of ideas must constantly be aware of the problems and questions which historical figures sought to solve.[10] Unfortunately, however, Collingwood's approach makes nonsense of a problem-solving historiography because of his idiosyncratic conception of problems and solutions. For instance, Collingwood is committed to the view that the *only* way the historian can determine what problems a thinker was trying to solve is by seeing what problems that thinker actually solved. As Collingwood says of Leibniz: "One and the same passage states his solution and serves as evidence of what the problem was. The fact that we can identify his problem is proof that he has solved it; for we only know what the problem is by arguing back from the solution."[11] On such an analysis, we could never say that a thinker failed to solve a problem because the only criterion Collingwood allows for attributing a problem to a thinker is that he has solved it. Such a Panglosian view of intellectual activity—entailing as it does that the only problems we ever attempt to solve are those which we actually solve—makes it impossible for the historian either to criticize the past or to explain its vicissitudes (at least in so far as the latter depend upon the failure of certain intellectual systems to solve the problems which they address). Collingwood failed to recognize that the historian can often find strong evidential grounds for attributing a problem to a thinker, even when that thinker fails to solve the problem which he has set for himself.

The Aims and Tools of Intellectual History

Chronology, exegesis, and explanation. Another of the core problems bedeviling the historiography of ideas is a crucial unclarity about the very aims of the enterprise. As construed by many of its practitioners, the aim of intellectual history is neither more nor less than *exegesis,* and its basic method is the classical one of *explication des textes.* Seen in this light, the primary task of the historian of ideas is to get clear about what people in the past have said and (in so far as he can get at it) what they have thought. One examines, for example, Newton's views on time or the Marxist theory of alienation and, in essence, tries to set out the appropriate doctrine in a clearer and more perspicacious way than its original proponent(s) did. Intellectual history, done this way, amounts to an elaborate form of paraphrase and précis. The historian sees his task as that of rehearsing the arguments he finds in the classical texts, perhaps filling in from time to time certain presuppositions which were not fully or carefully formulated in the original sources.

This sort of intellectual history I shall call *exegetical history,* precisely because its aim is straightforwardly explicative. Exegetical history aims at providing a *natural history of the mind as it evolves through time.* Like every other form of natural history, it is primarily *descriptive* in its ambitions. It seeks to record the temporal sequence of beliefs, in much the same fashion that descriptive geology aspires to record the sequence of changes on the face of the earth. But there is a very different type of intellectual history to which we might aspire, namely, *explanatory history.*

Our aim here would be not merely to rehearse what "great minds" have said but also to explain *why* they have said it. Clearly, exegetical history of ideas stands to explanatory history in the same relation as chronology stands to general history, or as any descriptive science stands to its explanatory counterpart. The explanatory scientist must be clear about the temporal succession of events, but he aspires to more than mere chronology. Indeed, he seeks to exhibit the reasons and causes that lie behind and explain the temporal sequences. In exactly the same way, the historian of ideas—if he intends to be more than a chronologist—must be ready to go beyond exegetical history. He must be prepared to ask, and to answer, questions like: Why did a certain thinker at a certain time espouse certain beliefs? Why was a given system of ideas modified when and where it was? How did one intellectual tradition or movement grow out of another?[12]

Unfortunately (and this may do much to explain the uneasiness many people have about it), scholarship in the field of intellectual history is still largely exegetical and not yet explanatory, neither in fact nor even in aspiration. The historiography of philosophy, almost certainly the most backward in this respect, provides some vivid examples:

For instance, scholars are widely agreed that the emergence of a hypothetico-deductive model of science was a very important characteristic of nineteenth-century logic and epistemology. Numerous exegetical studies have been written about the views of Kant, Whewell, Mill, Peirce, and others on that new philosophical model of science. Yet virtually no one has asked *why* it is the case that most nineteenth-century philosophers, unlike their eighteenth-century predecessors, thought it appropriate or important to stress the speculative nature of science. We have, as yet, not even the outlines of an explanatory history of epistemology and inductive logic for this period.[13]

Enlightenment historians of ideas have long agreed that the shadows of Bacon and Newton loom very large over eighteenth-century thought. Countless books and articles have been devoted to tracing out the influence of their ideas in France, Britain, and Germany during the period. Yet if one asks why Bacon and Newton were so much more influential than, say, Hobbes or Boyle or Malebranche, one finds that answers are neither frequently offered nor, when offered, cogently formulated. The fact of the dominance of Newton and Bacon in eighteenth-century thought has been documented *ad nauseam;* we have yet to make it a reasoned, or an explained, fact.

With individuals, as much as with broader movements, most intellectual history remains exegetical and nonexplanatory. It is now well known, for instance, that Newton and Leibniz were heavily influenced by Cartesian philosophy in their formative years. Yet both, for different reasons, came to repudiate Cartesian conceptions in their philosophical maturity. The chronology of this process has been well documented for some time. Yet if one asks for a convincing *explanation* of Newton's or of Leibniz's change of mind, contemporary scholarship has scarcely taken us beyond the sketchy explanations offered by Leibniz and Newton themselves.

The pervasive explanatory paucity of intellectual history, as illustrated by these few examples, is presumably more than accidental. There must be, one is inclined to conjecture, something about the current methods and presuppositions of the history of ideas which accounts for its explanatory bankruptcy. There are at least two areas where I am inclined to locate the difficulties: in the basic units of analysis hitherto utilized by historians of ideas; and in the difficulties that attend any effort to explain the beliefs of human agents. I shall deal with these in turn.

Concepts, "unit ideas," and research traditions. Until recently, the dominant mode of approach in the history of ideas has involved tracing one or more related ideas as they evolved through a long stretch of time. The concept of space, the idea of the great chain of being, the doctrine of *habeas corpus;* entities such as these have long been the stock-in-trade, the primary units of analysis, in intellectual history. This is hardly surprising; what else besides ideas would one expect the history of ideas to be concerned with? For all its initial plausibility, however, there is something profoundly deficient about focussing primarily on the concept, or (as Lovejoy called it) the "unit idea."

For one thing, such an approach tends to ignore the fact that ideas are interrelated and interconnected. If we want to understand what someone means by an idea, we must see how he uses it, how it functions for him in a broader framework of convictions about the world. In many cases, the very determination of the *meaning* of a concept or idea requires us to penetrate deeply the belief network of the thinker using the concept. As recent scholarship has shown, pinning down the meaning of (to cite a few examples) Newton's conception of matter, Faraday's conception of force, or Hobbes' conception of the state, requires an unpacking of the entire *Weltbild* of the thinker in question.

But there are other, even more serious ways in which the focus on single ideas puts acute obstacles in the way of historical analysis. As we know, ideas change and evolve. Accounting for such changes must be one of the central tasks of the history of ideas. Such changes can only be explained by looking at the shifting position of an idea within a broader conceptual network which is undergoing continuous modification. Hence, to explain the changes wrought on a particular and specific concept, we generally must look to a larger unit than the concept itself. Recent studies have shown, for instance, that the concept of "natural regularity," whose pre-history can be traced to antiquity, underwent a major alteration during the seventeenth century. We can begin to understand this change when we see that it was closely linked to the emergence of voluntarist theology and that the notion of a law of nature which comports with a freely acting deity is very different from the kind of natural regularity which one contemplates for a deterministically ordered, teleologically structured universe. To attempt to tell the history of the metaphysical concept of natural order, without at the same time telling the history of the larger systems or traditions of thought (embracing both science and theology) within which that concept was embedded, is evidently doomed to failure.

At a deeper level, the central danger with this approach to intellectual history is its tendency to blind historians to the changes wrought upon an idea or concept in the course of its evolution. To suggest, as Lovejoy does, that both Plato and Leibniz subscribe to the idea of the great chain of being is to gloss over the fact that "the chain of being" means something radically different for the two thinkers. To assert, as Holton does, that the "theme" of discontinuity is a recurrent one in human thought probably obscures more issues than it clarifies since (to take two extremes) Democritean discontinuities are very unlike the discontinuities postulated by Bohr or Planck.[14] What do we gain by seeing the history of thought as a counterpoint between such polarities as being and becoming, activity and passivity, or quantity and quality? Does it really explain what a thinker is doing to represent him as constructing a system by dipping into the familiar well of primordial concepts? My belief is that concepts evolve every bit as much as problems do and that the presumption of *stasis* at either level is tantamount to accepting an outmoded Platonic conception of the nature of intellectual history.

Recent philosophical, as well as historical, scholarship underscores the need to abandon the traditional "vertical" or unit idea approach to intellectual history. Thinkers like Duhem, Quine, Hanson, and Feyerabend have cogently argued that it is entire systems of thought which confront experience.[15] Individual concepts, particular propositions, which are components of these larger complexes, do not—indeed cannot—stand alone, and as a result we generally should not appraise or evaluate concepts on a piecemeal basis. Because these larger systems (which I have called "research traditions") function at any given time as the effective units of acceptance (or rejection), it follows that the intellectual historian—in so far as he wants to explain the evolving vicissitudes of belief—must take such traditions as his fundamental units for historical analysis.[15] This, in turn, requires a more *horizontal,* less vertical, approach than one customarily sees in historical scholarship. We must focus more on smaller time slices, wherein we examine the systematic interconnections between the concepts in several *contemporaneous* research traditions. If we would learn why Newton introduces the concept of absolute time, or why Locke modifies the traditional concept of monarchy, we must examine in detail their own traditions and the traditions of their rivals. We must be prepared to show how, for instance, the introduction of certain conceptual variations improved the overall problem-solving capacity of one or another system which embodies the variation.

There is another dimension to the contrast between "vertical" and "horizontal" history. It is sometimes argued, and even more widely assumed, that the central function of the history of ideas is to clarify what some "great thinker" meant in a given text. Dilthey's quest for *Verstehen,* Collingwood's preoccupation with "re-thinking another man's thoughts" and "getting inside his head," as well as Skinner's concern with identifying the "intentions" of great thinkers,[16] all speak of the preoccupation of theorists of intellectual history with faithful exegesis. Doubtless important, this process is, however, by no means the most important task of the intellectual historian. The historian must be at least as concerned with how ideas are received (what the Germans call *Rezeptionsgeschichte)* as he is with how they reached maturity in the head of the thinker who produced them in the first place. The intentions or the internal thought processes of a man who generates an idea are largely (and often completely) irrelevant to explaining how that idea is received in the appropriate intellectual community. Putting the point differently, if our primary focus is on the evolution of research traditions, then we need to pay relatively more attention to the ways in which the tradition is interpreted and modified by its defenders and relatively less attention than we have previously to the ratiocinative processes that produced the tradition in the first place.

In arguing that research traditions rather than individual concepts should be the basic unit of historical analysis, I am far from claiming that the intellectual historian should eschew ideas and concepts. My claim is, rather, that even if (perhaps especially if) we are interested in single concepts, we must begin with an analysis of research traditions, for it is the changing fortunes of the latter that generally serve to explain both the specific changes in, and the fortunes of, the former. We must not be misled by the fact that most physicists talk about space, or most political theorists talk about the state, into thinking that concepts like "space" and "the state" have an historical autonomy about them which allows one to explain their historical transformations independently of the broader patterns of belief of which the particular concept forms only one strand.

Explanations in intellectual history. If the failure to focus consistently on the most useful unit of analysis has caused some mischief, an even more serious problem has been an ambiguity about the explanatory ambitions of the history of ideas. In most of the explanatory sciences, the object of explanation, what is to be explained, is either an event (the falling of a stone), a process (the growth of a plant), or an action (the bombing of Hiroshima). By and large, the history of ideas is not primarily concerned with explaining any of the above. Its basic data are *beliefs,* and the changes and modifications of belief. If intellectual history is to be properly explanatory, its aim must be the explanation of the vicissitudes of belief and conviction on the part of historical agents. Merely documenting what those beliefs were and how they changed, which is the aim of exegetical history, clearly does not give us explanations. To secure the latter, we must offer cogent historical arguments which show why a given belief was formulated, accepted, modified, or rejected. It is here that the rub comes, however, for there is still much debate about what can legitimately count as "explaining a belief."

What should an *explanans* of the appropriate type look like? If we subscribe to the usual model of explanation, and apply it to the history of ideas, we might suggest that any adequate explanans must contain both some universal statements ("laws") and some statements of initial conditions. The two sets should jointly entail a proposition enunciating a belief situation that we wish to explain. Accepting this model for the moment, our question about explanation within intellectual history reduces to this: What sorts of things, if any, count as appropriate laws and initial conditions for the explanation of beliefs?

There are at least two different ways in which we might seek to answer this question. The first: if we were committed social (or psychological) determinists, convinced that all beliefs are caused by (and thus to be explained in terms of) the socio-economic position or psychological state of believers, then we should require laws relating a specific type of social situation, *x*, to a specific type of belief, *a* (namely, those occurring in the explanandum). Our initial conditions would (hopefully) assert that a particular believer *z* was in the relevant situation *x*. We could then deduce that (and thereby explain why) *z* accepted belief *a*. This type of explanation is rarely offered by historians of ideas; not surprisingly, for most of them do not subscribe to situational determinism of belief and thus are not willing to accept the truth of the "laws" invoked by this type of explanation. Because social accounts of belief are not a widely accepted mode of explanation, and because much of chapter seven is concerned with the ramifications of such an approach, I shall not discuss it further here.

The second: far more frequently invoked than the above are what we might call *rational explanations of belief*. Here we assume, implicitly or explicitly, certain rules or laws of rational belief and then apply them to particular belief situations. A historian might say (for instance) that Bacon rejected a belief in superstitious magic because he could see no evidence for it (assuming as his general explanatory law that "rational agents only accept beliefs when they have positive evidence for them"). Because this mode of explanation is so crucial for the very possibility of an explanatory history of ideas, it is worth examining its structure in more detail. Consider the following schema:

All rational agents in situation type, *a*, will accept (or reject, or modify) belief type, *b*. (1)
Smith was a rational agent. (2)
Smith was in situation a_1 (i.e., an *a* type situation). (3)

Smith accepted (or rejected, or modified) belief b_1. (4)

Statements (3) and (4) of schema (1) are presumably unproblematic; the evidence should establish their truth status unambiguously. Statement (2) is only slightly more difficult; sufficiently careful biographical studies can establish with high likelihood whether a given historical figure generally was or was not rational in his appraisal of beliefs in this field. By contrast, (1) is the problem case, for how do we discover laws or principles of type 1?

The question cannot be avoided or postponed, for a plausible answer to it is a necessary prerequisite to any history (as opposed to a chronology) of ideas. The general laws sought will belong, of course, to the theory of rational belief; for it is only such theories that can provide general principles of the type represented by statement (1). The applicability of such theories of rational belief, in turn, crucially depends on what we pack into our characterization of the believer's "situation type." As I pointed out in Part One, most theories of rational belief fail to be of much use to the historian because they deal with a very impoverished range of situation types.

On an *inductivist* theory of rationality, for example, the only situation types discussed would be those in which a belief was assigned a very high (or a very low) probability on the strength of the known empirical evidence. But, as we have seen, this is of little aid to either the historian of science or the general intellectual historian, because virtually no actual historical cases of belief exemplify the strict conditions demanded by inductivist models. In *deductivist* theories of rationality, on the other hand, the only allowable situation types would be those in which relations of entailment obtained between the belief to be explained and other beliefs of the agent. While such cases do certainly occur in the history of thought (and to this extent deductivist models of rationality have *more* to offer the intellectual historian than inductivist ones), they still constitute only a tiny proportion of the belief situations which he seeks to explain.

A variant of the deductivist model often invoked by intellectual historians is Collingwood's theory of presupposition. The idea here is to get at those core concepts which lie behind, as it were, the explicit beliefs to which a thinker subscribes. The problem is that presuppositional analysis (at least in its Collingwoodian form) is, at its core, purely deductivist. It can explain those beliefs of a historical figure which follow strictly from his alleged presuppositions; but it can explain neither the presuppositions themselves nor any beliefs which fail to be deductive consequences of those presuppositions. Still worse, presuppositional history offers no machinery for discussing why historical agents accepted one set of presuppositions rather than another. Thus it leaves unexplained that very aspect of history which it takes to be the most important.

Apart from the limitations already discussed, these models of rational belief suffer from yet another liability when applied to the history of ideas: namely, their insensitivity to (tantamount to a denial of) the degree to which *specific canons of rationality are time-dependent.* A mode of argument which one epoch, or "school of thought," views as entirely legitimate and reasonable may be viewed by another era or another intellectual tradition as ill-founded and obscurantist. Neither inductivist nor deductivist theories of rationality leaves the historian any scope for attending to those subtle, temporal shifts in standards of argumentation which continuously confront him in his research.

What intellectual history stands most in need of, in my view, is a theory of rational belief which goes beyond the restrictive limitations of the inductivist and deductivist models.

The problem-solving model of rationality discussed earlier represents a step in that direction. It is sensitive to shifting, local canons of rational belief; it does allow for a comparative, rational appraisal of presuppositions; it does not limit rational belief to those cases in which there is a rigid deductive or inductive propositional link.

Such grandiose claims may sound fine in the abstract, but how in practice does the problem-solving model shed any light on specific cases? The method of application of the model is relatively straightforward. One begins by identifying the pool of available explanatory systems (i.e., research traditions) in any given epoch and intellectual community. One then determines for each of these research traditions how progressive they were (i.e., how effective they were at maximizing solved problems and at minimizing anomalous and conceptual ones). This analysis will allow the historian to construct a profile on the progress of each of the available options.[17] Independently of these profiles, one has a number of laws or general principles of rationality. Among them would be principles such as: (1) all rational agents will prefer a more effective research tradition to a less effective one; and (2) all rational agents, in modifying a research tradition, will prefer more progressive to less progressive modifications of it.

Such principles, when conjoined with the progress or rationality profiles of each of the available research traditions, will allow one to *explain* many developments within the history of thought which have thus far eluded explanation. Such, at any rate, is the claim being made for the problem-solving model.

It might be held that, in order to give historical explanations of the kind proposed here, we have no need for rational, normative appraisals whatever. It could be said that it is not the historian's task to determine whether some belief was rational, only to show that some thinker *thought it to be so.* Suppose, for instance, we wish to explain why Newton advocated action-at-a-distance forces to explain gravitation. Is it not enough merely to recite Newton's stated reasons for introducing the concept, adding perhaps that he regarded these as sufficient reasons for using the concept? On this analysis, there is no place for the historian to ask the normative question whether, by the then appropriate canons of scientific belief, Newton *was right* in judging action-at-a-distance to be well conceived.

To locate the flaw in this approach, we may consider a second, parallel example. Suppose we wish to explain why a certain "special creationist" believes there was a universal deluge during the time of Noah. Suppose, further, that we can show that his only reason for this belief is that it accords with Scriptures and since he takes Scriptural concordance as a sign of truth, thinks his belief is well-founded. Confronted with such an "explanation," we would feel that the historian's job was only half completed, for we now want to know why this creationist subscribed to such a peculiar theory of truth. Our curiosity is only whetted, not satisfied, by being told that someone accepted a belief for which there were only "bad" reasons, not "good" ones.

By contrast, if we can show that a thinker accepted a certain belief which was really the best available in the situation, then we feel that our explanatory task is over. Implicit in this way of looking at the matter is the assumption that *when a thinker does what it is rational to do, we need inquire no further into the causes of his action;* whereas, when he does what is in fact irrational—even if he believes it to be rational—we require some further explanation. This assumption thus functions in the realm of human behavior very like the principle of inertia within mechanics. In both cases, the principles provide a characterization of what we regard as "normal behavior." A body moving at constant velocity and a man behaving rationally are both "expected states," which require no further causal analysis. It is only when bodies change velocity or when men act irrationally that we require an explanation of these deviations from the expected pattern. Of course, this proposal—that rational behavior is the rule rather than the exception—is open to debate, but as we shall see in chapter seven, it is preferable to the alternatives. Precisely because it is preferable, normative evaluations—as opposed to purely descriptive ones—must play a role in historical explanations, for those evaluations tell us when our explanatory task is at an end.

Problem Solving and Nonscientific Research Traditions

It might be thought that the problem-solving model articulated in Part One, although applicable to the history of scientific ideas, is of only very limited use in those areas of intellectual history which deal with nonscientific domains. While conceptual problems obviously occur in all fields of inquiry, empirical problems seem to occur far less globally. It has, after all, been argued at length by many scholars that the sciences alone are empirical disciplines, from which it would follow that science alone has what I have called empirical problems and that there would be no counterpart of empirical problem solving in the nonscientific disciplines. If it were true that the natural and social sciences exhausted the range of empirical problems (as, for instance, the positivists maintained) then one would have grave doubts about the suitability of a problem-solving model for dealing with general intellectual history. But to imagine that "nonscientific" disciplines traditionally have had no significant empirical element is a gross historical travesty. Consider only a few examples:

1. Metaphysics is frequently cited (especially by professional anti-metaphysicians) as an ideal example of a discipline with no

empirical content. But there are, and classically have been, a host of empirical problems which metaphysical systems have sought to resolve. For instance, most objects are seen in our daily experience to endure through time. One of the core empirical problems of metaphysics has been explaining what properties of being can explain the seeming endurance of objects. Similarly, most changes which we experience in the world seem to be causally linked to other changes. An exploration of the causal nexus has been a persistent problem for metaphysics. Even those metaphysical systems (like occasionalism) which deny the ultimate reality of a causal connection between events, still must explain an empirical problem; to wit, why it is that the world seems to be causally interconnected. It is doubtlessly true that the specificity of the empirical problems which confront, say, the chemist and the ontologist are very different; but the difference is one of degree, not kind. The metaphysician and the historian of metaphysics, every bit as much as the chemist and his historian, must attend to the empirical problems of the field.

2. Theology, like metaphysics, is often alleged to be empirically transcendent and thus devoid of empirical problems. But few traditional theologians or historians of theology would subscribe to such a view. For instance, the "problem of evil" is at its core an empirical problem *par excellence:* how can one maintain one's belief in a benevolent, omnipotent deity in the face of all of the death, disease, and natural disasters which are a daily element of our experience? Many theological doctrines have been devised largely to deal with this seeming empirical anomaly. Judeo-Christian theology is, more than most, replete with a large body of similar empirical problems. On one level, that theology makes certain historical claims about the existence of persons and the occurrence of events. At another level, Judeo-Christian theology makes claims about the experiential effects of "true belief" on believers. These claims are, in principle, testable within the realm of experience.[18] If false, these claims are confronted by a large base of empirical anomalies which any adequate (i.e., progressive) theology must either solve, or suffer the cognitive consequences of failing to solve. If true, then they constitute solved empirical problems.

Similar remarks could be made about the existence of empirical problems in every other branch of human inquiry. Even in the so-called formal sciences, such as logic and mathematics, where one might least anticipate empirical problems, they exist in large numbers—as Lakatos' fascinating studies on the history of mathematics amply demonstrate.[19]

The applicability of a problem-solving model to nonscientific disciplines has implications not only for writing the history of such disciples, but also *for appraising their cognitive status.* It is frequently claimed that the sciences alone are progressive and cumulative, while other areas of inquiry exhibit changes of fashion and style which cannot be meaningfully described as progressive.[20] The contrast is sometimes put differently; it is sometimes said that the sciences can discover when their assumptions are wrong, but the humanistic disciplines cannot; it is frequently alleged that the sciences are "self-corrective," but that the nonsciences lack that crucial characteristic. However the distinction is put (progressive v. non-progressive, rational v. non-rational, empirical v. non-empirical, falsifiable v. non-falsifiable), it will not hold up to detailed scrutiny. Disciplines like metaphysics, theology, even literary criticism exhibit all the features we require for making rational appraisals of the relative merits of competing ideologies within them. The nonsciences, every bit as much as the sciences, have empirical and conceptual problems; both have criteria for assessing the adequacy of solutions to problems; both can be shown to have made significant progress at certain stages of their historical evolution.

What has stood in the way of a recognition of the cognitive parity of the sciences and the nonsciences has been a simplistic identification of (scientific) rationality with experimental control and quantitative precision. Because "humanistic" theories usually lack both, it has been easy for some thinkers to dismiss their rational credentials. But, as we have seen, the essence of rationality in science does not depend on such characteristics.

That much said, however, we ought not jump to the opposite extreme. That it is *possible* and *appropriate* to talk of progress and rationality in the nonsciences I take as established, at least programatically; it clearly does not follow, however, that the various humanistic disciplines have *in fact* been as progressive and as rational as the sciences. Progress, as we said in Part One, is a matter of degree; two systems of thought can each be progressive, while one may show a higher rate of progress than another.

If there is any truth at all in the (positivistic) claim about the differences between the sciences and the nonsciences, and I suspect there is some truth in it, it will be found, not in the exclusive exhibition of progress by the sciences, but rather in the higher rate of progress exhibited by them. But even this claim is still a matter of vague intuition, and will remain so until historians of nonscientific ideas begin re-writing history with a view toward appraising the relative progress and rationality of competing research traditions in the humanities.

There is one final aspect of the humanities-sciences contrast which requires comment. It is frequently alleged that the adoption of doctrines in the nonsciences can only be a subjective matter of taste and fashion. If one becomes an empiricist, or idealist, or a trinitarian, or a socialist, the decision (so it is claimed) is entirely arbitrary. None of the positions can be "proven" true or false, and there are always arguments pro and con. As a piece of descriptive social psychology, there is doubtless much to be said for this view. Many persons do indeed see (and make) the choice between competing ideologies as an intrinsically nonrational affair. But there is no reason in principle why this need be the case. The choice between atheism and theism, between phenomenalism and realism, between intuitionism and formalism, between capitalism and socialism (to cite only a few examples) could be made by appraising the relative progressiveness (and thereby the relative rationality) of these competing research traditions. If we could show (as I suspect we can for all the pairs cited above) that, at the moment, one tradition has been a more progressive problem solver than its competitors, then we would have legitimate, rational grounds for preferring it. If and only if the competing traditions emerged from the analysis with equally progressive evaluations would we then be entitled to argue that the choice between them was necessarily arbitrary and conventional. The presumption that the acceptance or rejection of ideologies can never in principle be rationally justified (a presumption at the core of sociology of knowledge) is, on this analysis, entirely unfounded.

The Indispensability of History for Theory Appraisal

Thus far in this chapter, our concern has been to examine a few of the foundational problems in the historiography of ideas. By way of conclusion, however, I want to shift the discussion towards a consideration of the *relevance* of intellectual history to *contemporary* cases of theory evaluation. It has often been argued that any attempt to utilize the historical evolution of a system of ideas as a vehicle for criticizing or appraising its current status is a category mistake. Logicians teach us that it is a special version of the so-called genetic fallacy to imagine that the origin or historical career of a doctrine has anything whatever to do with its cognitive well-foundedness. Offering us a sophisticated version of Henry Ford's dictum that "history is bunk," most modern theories of rational appraisal insist that the temporal career of a doctrine or research tradition is absolutely irrelevant to its rational acceptability.[21] I want to take exception with this view, even to turn it on its head, by arguing that *no sensible rational appraisal can be made of any doctrine without a rich knowledge of its historical development* (and of the history of its rivals).

What leads to these very divergent perspectives on the relevance of intellectual history is a deep disagreement about the aims and nature of rational appraisal itself. If one takes the traditional view that, in appraising any doctrine, we should identify rational belief with truth presumptiveness, then the history of any doctrine is indeed largely irrelevant to its rational status. A doctrine could be imagined to have virtually any prior history whatever and yet still be true; similarly, a false doctrine could conceivably exhibit any historical pattern one likes. The crucial trouble here, of course, is that—for reasons already discussed—we have no way of determining whether a (consistent) system or theory is true or false, or even presumptively one or the other. As a result, appraisals of the rationality of accepting any doctrine must be based on factors other than their truth status. I suggested earlier that the most promising factor on which to pin our acceptance is "progress at problem solution."

If, however, we once accept the proposal that the appraisal of any doctrine should be based on the problem-solving progressiveness and effectiveness of the research tradition with which it is associated, then we are inevitably committed to the view that intellectual history must be an *ineliminable* ingredient in every rational choice situation. For until we know how a research tradition has fared through time (especially relative to its extant rivals) we are in no position to appraise its rational credentials. To a certain degree, such an approach as that proposed here is already widely used. The claim that "logical positivism has run out of steam," the observation that "the New Criticism is no longer a promising device for literary analysis," the charge that "psychoanalysis is becoming increasingly ad hoc and doctrinaire"; these, and similar, familiar characterizations already exploit the insight that a tradition's history is relevant to an appraisal of its current cognitive status.

But this mode of analysis is still sufficiently undeveloped that it has been assumed that a very superficial historical "intuition" about how a tradition has evolved is enough. If, however, we take this view as seriously as it deserves then we need far more than vague impressions about the temporal dimensions of a research tradition. What we need, if our appraisals are to be at all reliable, is serious historical scholarship devoted to the various research traditions in any given field of inquiry. Without the information which such studies can generate, it is impossible to make an informed and rational choice between competing ideologies within any domain. In this sense, and to this degree, all contemporary disciplines are, or ought to be, parasitic upon their intellectual ancestry—not only genetically but cognitively.

This last point puts us in a position to return to the objections voiced by general historians about intellectual history as an enterprise. Those objections, in so far as they imply that general history can dispense with the history of ideas, must fall wide of the mark if the arguments of this chapter are cogent. For history itself is a theoretical discipline with rival ideologies, alternative methodologies, and competing traditions; sensible choice between those traditions hinges, as we have seen, on an awareness of the intellectual history of those ideologies. Hence, for all its alleged "elitism" and "idealism," *intellectual history,* far from being at the periphery of the concerns of the general historian, is directly at the core of any historical research, and *is presupposed by every other form of history*—at least to the extent that the general historian's problems and methodologies do themselves have an intellectual history of which the historian must be aware if he is to write sound history.

But to say as much is only to insist that the social or economic historian must be aware of the intellectual history of history itself. We have not yet challenged the common claim that the history of ideas must be replaced by a broader form of socio-economic history whose function would be the identification of the "real," nonintellectual causes of changing patterns of belief. It is that specific issue which we must now confront.

Chapter Seven

Rationality and the Sociology of Knowledge

A man always has two reasons
for doing anything—a good reason
and the real reason. J. PIERPONT MORGAN

Anyone who wants to drag in
the irrational where the
lucidity and acuity of reason
still must rule by right
merely shows that he is
afraid to face the mystery
at its legitimate place. KARL MANNHEIM (1952), p. 229

One of the most important controversies within the community of scholars who study the evolution of science has concerned the role of sociological and psychological factors in the development of scientific thought. It is at this nexus between the "internal" and the "external" that intellectual historians of science cross swords with social historians of science, and where those who favor a rational analysis of science quarrel with historical sociologists and psycho-historians of science. Recently, this controversy has generated more heat than light, which is unfortunate because it is a genuine controversy, the outcome of which may do much to shape our general conception of science itself. There is, of course, a huge and burgeoning literature in the sociology of science. The object of this chapter is not primarily to discuss the detailed conclusions now emerging within that field; its aim, rather, is to examine the explanatory scope and range of the sociology of scientific knowledge in particular, and the sociology of knowledge (of which the former is a part) in general.[1] I shall try to show in particular that the model of rationality outlined in Part One has many ramifications for an understanding of the nature and limits of sociology of knowledge.

We must begin, however, with some preliminary distinctions, for much of the confusion in this area has arisen from a failure to bear in mind some elementary *differentia.* It is vital to distinguish, at the outset, between two very different sorts of sociologies of science: (1) suppose someone wishes to explain why a particular scientific society or institution was founded, why a scientist's reputation waned, why a particular laboratory was established when and where it was, or why the number of German scientists rose dramatically between 1820 and 1860. I propose to call the investigation of such problems *the non-cognitive sociology of science.* Such studies are non-cognitive precisely because their primary concern is not to explain the *beliefs* of scientists about the natural world, but rather to explain their modes of organization and their institutional structures. (It is true, of course, that scientists' beliefs may condition their modes of institutional organization;[2] but what makes this form of sociology non-cognitive is that the problems it sets itself to solve are not themselves beliefs about the natural world.) (2) By contrast, a sociologist may seek to explain why a certain *theory* was discovered (or, after discovery, accepted or rejected) by pointing to the social or economic factors that predisposed scientists to be sympathetic or hostile to it. Alternatively, he may seek to show that certain social structures were influential on the genesis of the concepts of a theory. Such efforts fall within the scope of what I shall call *the cognitive sociology of science.* Clearly, these two modes of approach, the cognitive and the non-cognitive, could be applied to any intellectual discipline, ranging from the specific sciences to theology, metaphysics, or sociology itself. As a result, we can speak more generally of the non-cognitive and the cognitive sociology of knowledge.

From what was said in chapter six, it should be clear that there is neither overlap nor conflict between the intellectual historian of science (or knowledge) and the *non-cognitive* sociologist, because they are addressing themselves to radically different problem situations. The intellectual historian is trying to explain why scientists or other thinkers in the past adopted the beliefs or the solutions (theories) which they did; the non-cognitive sociologist does not, by definition, have beliefs about the world among his class of problems to be solved. Quite the reverse is the case, however, when we compare cognitive sociology of science with the intellectual or rational historiography of science. For here, there is the possibility for enormous (and potentially fruitful) conflict. The intellectual historian of knowledge will generally seek to explain why some agent believed some theory by talking about the arguments and the evidence for and against the theory and its competitors. The cognitive sociologist of knowledge, on the other hand, will generally try to explain why the agent believed the theory in terms of the social, economic, psychological, and institutional circumstances in which the agent found himself. Both are trying to solve the same problem (namely, the belief of some historical agent), yet their modes of solution are so different as to be almost incommensurable. Is there any way, given these conflicting explanatory strategies, of determining who is right, the intellectual historian or the cognitive sociologist? Or could they *both* be?

The possibility of answering this important question hinges upon whether we can articulate any fair criteria for deciding between the seemingly competing historical accounts given by the cognitive sociologist and the intellectual historian. The articulation of such criteria is one of the central aims of this chapter.

The Domain of Cognitive Sociology

Before we proceed to that task, however, we must get clearer about the character of cognitive sociology, for some of its ablest practitioners seem at times to have made confusing, even inconsistent, claims about the scope of sociological theory and about the nature of sociological explanations.

The Nature of Cognitive Sociology

As we have already seen, one important feature of cognitive sociology is that it takes *beliefs* to be its empirical problems. But that is clearly not enough to distinguish it from many other nonsociological modes of explaining belief (such as, for instance, the rational history of science). What further distinguishes cognitive sociology from these other fields, what makes it *socio*logical, must be an assumption that beliefs are to be explained in *terms of the social situations of believers.* So we can say that an essential task of any cognitive sociologist must be that of exhibiting, for any belief he wishes to explain, its social roots and origins. To say as much is only to characterize what a sociological explanation will look like *once we have formulated it.* But presumably what we need as well is some way of identifying those belief situations which are likely to be amenable to a sociological analysis.

There are those who would maintain that virtually every shift in belief in any community of thinkers is explicable in terms of the social substructures, and would thus make the problem domain of the sociology of knowledge co-extensive with the entire history of human thought.[3] At the other extreme are certain critics of the sociology of knowledge who have claimed that there are virtually no transformations in the history of ideas which are in any way indebted to, or functions of, changes in social structure. The uncompromising social determinists (e.g., certain Marxists—though not Marx himself) and the intransigent idealists (e.g., Hegel) respectively exemplify these two poles.[4] Unfortunately, neither point of view makes much sense of the historical record. There is an enormous amount of evidence which shows that certain doctrines and ideas bear no straightforward relation to the exigencies of social circumstance: to cite but two examples, the principle that "2 + 2 = 4" or the idea that "most heavy bodies fall downwards when released" are beliefs to which persons from a wide variety of cultural and social situations subscribe. Anyone who would suggest that such beliefs were socially determined or conditioned would betray a remarkable ignorance of the ways in which such beliefs were generated and established. Similarly, there clearly are ideas and beliefs which do have tangible social roots and origins. To imagine, for instance, that a nineteenth-century white slave-holder espouses a belief in the racial inferiority of blacks for purely intellectual reasons requires a feat of moral charity which few of us could make. To suggest that most nineteenth-century German factory workers who favored socialism did so because of the rational well-foundedness of its doctrines is again a view that demands an enormous degree of credulity.

But if we grant that the truth of the matter lies somewhere between rigid social determination and insular idealism, we are immediately confronted with a central problem, to wit, *what sorts of beliefs are candidates for a sociological analysis and which are not?* Put differently, and in the language of earlier chapters, what kinds of belief situations can function as legitimate empirical problems for sociology? One might think that this is a purely empirical question which cannot be settled in advance *a priori,* but which can only be decided by looking at matters case-by-case. The troubles with this seemingly innocuous answer to our question are both practical and theoretical. On the practical side, we run up against the fact that there are, within the extant record, literally millions of beliefs. If the sociologist has no regulative principles which will guide his initial selection of potentially promising problems, he could make scarcely any headway. For example, one might ask about *each* of the truths of arithmetic whether they have social origins. We should begin with, say, "$1 + 1 = 2$" and work up the scale.

Because of such practical difficulties with a purely empirical approach to specifying the problem-set for cognitive sociology, virtually all researchers in the field have sought to delimit the domain of possible sociological problems by adopting certain regulative or methodological principles, whose function is to provide a useful initial sorting mechanism that will draw attention to those types of beliefs which are most likely to be susceptible to sociological analysis.

But there are theoretical, as well as practical, reasons for laying down in advance some way of deciding on the boundaries of the potential problems for the cognitive sociologist. If it were true that *all* beliefs were not the result of rational deliberation or enlightened evaluation, but rather were simply determined by the social situation of the believer, then the whole enterprise of cognitive sociology would be self-indicting; for if *all* beliefs are socially caused, rather than rationally well-founded, then the beliefs of the cognitive sociologist himself have no relevant rational credentials and hence no special claims to acceptability.[5] Ernst Grünwald put the point tellingly when he remarked: "For the thesis that all thought is existentially [i.e., socially] determined and thus cannot claim to be true does itself claim to be true."[6] Thus, the cognitive sociologist, to avoid being hoisted with his own petard, is committed to the view that *some* beliefs are rationally well-founded rather than socially determined.

There are three different methodological principles most often cited (or used implicitly) in this connection by cognitive sociologists of knowledge. I shall call them *the arationality assumption, the historico-social assumption,* and *the interdisciplinary assumption.* Although not strictly compatible, these conditions are widely (and often simultaneously) used in most works on the sociology of knowledge. I want to discuss them in some detail, for the model of science and knowledge developed in Part One of this essay impinges upon each of them and thereby upon the whole of cognitive sociology of knowledge.

The Arationality Assumption. Many sociologists of knowledge, following Karl Mannheim, distinguish between "immanent" and "non-immanent" (or "existentially determined") ideas.[7] *Immanent ideas* (or concepts or propositions or beliefs—these are all run together by most writers) are those which can be shown to be naturally and rationally linked to other ideas to which a believer adheres. An archetypal example would be the theorems of Euclid's geometry. Once one accepts the axioms, one is logically or rationally constrained to accept their theorematic consequences. No thinking person who understood the one, could deny the other. *Non-immanent* (existential) *ideas,* on the other hand, are those which do not carry their rational credentials with them. They are ideas which people may accept, but which are not intrinsically more rational than many other alternative ideas which they might have accepted.

Most sociologists of knowledge agree with Mannheim that *it is only non-immanent ideas,* only those which are not the most rationally well-founded in a given situation, *which it is appropriate for sociology to attempt to explain.* It is easy to see the plausibility of this stipulation. If the acceptance of some belief, *x,* seems to follow naturally and rationally from the prior acceptance of beliefs *y* and *z,* then there seems no point in maintaining that the espousal of *x* is directly caused by social or economic circumstances.[8] If, on the other hand, someone accepts a belief *a,* which is not rationally related to his other beliefs *b, c, . . . , i,* then it seems as if the only natural way of explaining his espousal of *a* will be in terms of factors that are extra-rational, such as the social (or psychological) situation of the believer in question.

I propose to call this demarcation criterion *the arationality assumption;* basically, it amounts to the claim that *the sociology of knowledge may step in to explain beliefs if and only if those beliefs cannot be explained in terms of their rational merits.* As Robert Merton points out, this view is widely accepted by working sociologists: "A central point of agreement in all approaches to the sociology of knowledge is the thesis that thought has an existential [i.e., social] basis in so far as it is not immanently [i.e., rationally] determined."[9] Essentially, the arationality assumption establishes a division of labor between the historian of ideas and the sociologist of knowledge; saying, in effect, that the historian of ideas, using the machinery available to him, can explain the history of thought insofar as it is rationally well-founded and that the sociologist of knowledge steps in at precisely those points where a rational analysis of the acceptance (or rejection) of an idea fails to square with the actual situation.

The arationality assumption, we must stress, is a *methodological* principle, not a metaphysical doctrine. It does not assert that "whenever a belief can be explained by adequate reasons, then it could not have been socially caused"; it makes the weaker, programmatic proposal that "whenever a belief can be explained by adequate reasons, there is no need for, and little promise in, seeking out an alternative explanation in terms of social causes."

Although the arationality assumption is widely subscribed to by cognitive sociologists, there are few arguments ever cited for its cogency. Because it has recently come under attack by historical sociologists and because it is so crucial as a demarcation criterion between rational explanations of belief and extra-rational explanations thereof, it is worth exploring briefly the grounds for it. In order to do so, let us suppose the following imaginary situation: there is some person, *x*, who believes *A*. His belief patterns are being investigated by two researchers, *y* and *z*. Suppose *y* is an intellectual historian who takes the arationality assumption seriously; he looks for, and finds, a way of showing that *x*'s belief is rationally well-founded, given *x*'s other beliefs *B, C, . . . , I*. So far as *y* is concerned, he now has as full an explanation of *x*'s belief in *A* as seems possible. Suppose, however, that *z* is a maverick sociologist who refuses to accept the arationality assumption. While granting that *y* has found a "rational" explanation of *x*'s belief, *z* is convinced that there may still be scope for sociological work on belief *A* (perhaps because *z* suspects that *y* has mistaken a "rationalization" of *x*'s for the "real" cause for *x*'s espousing *A*). After some biographical research on *x*, *z* discovers that *x* was a member of a lower middle class and had an Oedipal fixation on his mother. Suppose, further, that *z* would argue that persons in *x*'s situation generally tend to hold beliefs such as *A*. While not denying that *y* has offered an alternative explanation of *x*'s belief, the sociologist *z* nonetheless insists that his own explanation still stands; that it is, if anything, "more fundamental" than *y*'s explanation. How, if at all, can *y* convince *z* that his explanation is bogus because it violates the arationality assumption?

One might, of course, simply postulate the arationality assumption as a matter of faith; as a postulate without which it would perhaps never be possible to choose between conflicting accounts of human belief. But such pious hand waving will scarcely convince our resolute social determinist, *z*. What may help is an analysis of *z*'s own intellectual orientation. *Z*, and like-minded types, are attempting to explain beliefs. Any explanation, if it is cogent, is an argument, a reasoning process that moves from adequate premises to plausible conclusions. The whole point of offering an explanation, unless it be an idle speech act, is to demonstrate that the conclusion follows rationally from the premises. So that *z*, insofar as he offers sociological explanations, is presuming that at least some persons (notably himself) accept certain beliefs because they have good reasons for doing so. (One presumes here that *z* would not take kindly to the suggestion that the only cause of his belief in a certain sociological explanation was his location in the social network!) But if *z* insists that certain agents' beliefs (namely, his own) are rationally well-founded and not merely a function of their social position, then the onus is on him to show why it is appropriate to regard his own beliefs as situationally transcendent, while the beliefs of the persons he studies—even when they can be rationally explained—ought not be viewed as independent of their social situation.

There is a very different manner in which one might seek to adjudicate this controversy between *y* and *z*, namely, by viewing their theoretical systems (in the language of Part One) *as competing research traditions.* Approached in that way, we could ask which has solved more important empirical problems. There can be no doubt whatever that, at least at this point in time, the rational historiography of ideas has gone much further toward explaining a large number of important historical cases of belief than historical sociology has. Indeed, the "success ratio" of intellectual history is several orders of magnitude greater than that of cognitive sociology.[10] At the level of conceptual problems, too, the traditions of the intellectual historian are generally acknowledged to be in less acute difficulty than those of the cognitive sociologist.[11] Under such circumstances it would be entirely appropriate to point out to *z* that, whenever we have competing rational and sociological explanations *of the same belief,* good sense dictates that the "rational" explanation should be given priority over the sociological one precisely because the former has shown itself to be the most fruitful. (Which is not to say, of course, that sociological explanations are inappropriate ***where rational reconstructions fail to apply.)***

Whether it is for reasons such as these that most cognitive sociologists adhere to the arationality assumption, I do not know. But whatever its grounds might be, it is taken as axiomatic by most practitioners, and it is important for us here to examine some of its important consequences.

Despite its pervasiveness, it has been little noticed that the arationality assumption is a great deal more problematic than most of its proponents recognize. In order to apply it, we obviously need a theory about what rational belief is. Without such a theory, the arationality condition is meaningless. But as we have seen in Part One, and as should have been clear all along, there is more than one conceivable theory of rationality. Because different theories of rationality will classify beliefs differently (certain theories making a particular belief rational while other theories make the same belief irrational), we can see that ***an essential prolegomenon to any adequate cognitive sociology of knowledge is the choice as to a theory of rationality.***[12] If we accept, as some working sociologists are prone to, a simple-minded theory of rationality which puts excessive constraints on what is to count as rational belief, then the domain of the arational—and thus the domain of the sociological—is going to loom very large. If, on the other hand, we subscribe to a richer theory of rationality, far more beliefs are going to seem "immanent" and thus not susceptible to sociological analysis.

Insensitivity to the variety of theories about rational belief has been the source of much mischief and confusion in the writings of many prominent sociologists. Assuming that the "textbook, inductivist" theory of rationality which they inherited from philosophers of science was sacrosanct and final, sociologists have been prone to regard as irrational (and thus as sociological) many episodes in the history of thought which are, by other standards of rationality, entirely rational. This, in turn, has led sociologists to seek social causes for processes which can be explained entirely in immanentist terms.

If, for instance, we subscribe to a crude "empiricist" model of rationality according to which the empirical success of a theory is the only relevant determinant of its rational acceptability, we are going to look askance on those episodes in the history of thought where (to use the language of Part One) conceptual problems play a major role in determining which theories will be accepted or rejected. If a theory in the past has been objected to on the grounds of its incompatibility with a certain metaphysical or epistemological or theological belief structure, proponents of this limited, empirical model of rationality will see the episode as intrinsically irrational, as one in which certain ill-founded prejudices were allowed to controvert the rational judgments of the agents in question. This, in turn, will lead to the conclusion that social factors must have had something to do with the outcome of the decision, for the rational canons of preference were seemingly ignored.

What vitiates this approach to history, of course, is the existence of other models of rational belief which would make it perfectly reasonable, under *certain* circumstances, for factors of a philosophical or theological kind to enter into the rational appraisal of a particular theory. Viewed through the perspective of such models, developments which might previously have been regarded as prejudiced, obscurantist and irrational acquire a rational legitimacy, which obviates the need to look to the social milieu for an explanation of what is going on. The moral should be clear: before we classify an episode as arational, before we begin the search for social causes to explain the "deviations" from the rational norm, we must be quite sure that our notion of rationality is an adequate one. To my knowledge, few if any sociologists have seen the force of this point, and their work is the worse for their failure to see it. Unfortunately, the error is doubly confounded; in addition to their failure to recognize that there may be a broad spectrum of theories of rationality, they have also generally chosen to subscribe to that model of rationality which is the most limited of all.

To see just how pervasive this error is, it may be useful to look at a few prominent examples. In his influential *Structure of Scientific Revolutions,* Thomas Kuhn considers several of the better-known "empiricist" models of scientific rationality which philosophers have espoused. He finds both the confirmational and the falsificational models inadequate, but goes on from there to enunciate his own model of scientific rationality. In its essential features, this model is a purely empiricist one which shares with the other models the conviction that it is only the empirical problem-solving ability of a theory which can be relevant to its rational appraisal. Kuhn then points out, quite rightly, that there are many episodes in the history of science which seem to involve decisions about theories where factors other than the empirical credentials of the theories under examination were prominent.[13] Kuhn argues, or rather alleges without argument, that in such cases there must be important social and institutional pressures at work. In making this slide, Kuhn is obviously, if implicitly, invoking the arationality condition. There can be no objection to that, but one wishes he might have worried more deeply about what rationality amounts to before jumping to the conclusion that his empirical model of rationality was subtle enough to provide the careful discriminations between the immanent and arational.

Similar precipitate leaps to the presumption of arationality occur repeatedly in Maurice Richter's recent *Science as a Cultural Process.* Richter argues, for instance, that Darwin's theory of evolution: "was challenged in the nineteenth century not only on the basis of *reasonable scientific arguments* . . . but also on the basis of dogmatic theological assumptions."[14] Richter may, of course, be correct in his historical claim; but the image of scientific rationality which forms the backdrop to his notion of "reasonable scientific arguments" is suspect, at best. He maintains, for example, that "the contents of scientific knowledge . . . are to be determined by observations of nature."[15] This highly empiricist notion of what constitutes proper science leads Richter, not surprisingly, to see many historical episodes as irrational (because not reconstructible on a naive empiricist model of rationality) and thus as sociological.

One of the most striking instances of the hyper-positivistic current in the cognitive sociology of science is provided by the work of the well-known sociologist Bernard Barber. In a widely cited article in *Science* in 1961,[16] Barber explored the various factors which dispose scientists to refuse to accept new ideas and new discoveries. In this latter-day version of Bacon's "idols," Barber identifies methodology and theology as two of the major sources of "cultural resistance to new ideas." There is clearly nothing wrong with Barber's hunch that philosophy and theology have played an important role in scientific debate. His positivism only comes out when, having noted the interaction, he goes on to bemoan it and to urge that we should seek to reduce their pernicious influence.[17] Barber has recognized neither that it is often perfectly reasonable, and not just prejudice, to attend to the broader methodological and philosophical implications of a new scientific theory, nor that methodology and theology have historically been involved as often to legitimate new theories as they have been called upon to discredit them. Barber yearns for what he calls the "open-minded" scientist, who restricts himself entirely to the straightforward, "scientific" merits of a new idea. Barber's purely empiricist model of theory appraisal allows no room for anything else.

In these, as in numerous other cases that could be cited from the recent literature, scholars seem to have leapt prematurely to the conclusion that the inapplicability of one or other standard model of rationality to any particular case establishes the arationality (and thereby the social character) of the case at hand. It should be clear that if we accept a different model of rationality, one perhaps built along the lines sketched earlier in this essay, then the domain of would-be sociological cases will be far smaller than if we accept one of the more traditional empirical theories of rationality. (My own proposal would be that the need for sociological analysis of a case only arises when we can show that the actual evaluation of a particular theory in the past was radically at odds with the appraisal it should be accorded by the lights of the problem-solving model of rationality.)

I have dwelt at length on the parasitic dependence of cognitive sociology of knowledge on theories of rationality not only to draw attention to the need for sociologists to be more self-critical about their judgments concerning the rationality of particular cases, but also to emphasize the fact that *the application of cognitive sociology to historical cases must await the prior results of the application of the methods of intellectual history to those cases.* The cognitive sociologist must look to the intellectual historian for cues and clues about which cases it is appropriate for him to analyze. Until the rational history of any episode has been written (and that, by using the best available theory of rationality), the cognitive sociologist must simply bide his time; to do otherwise is to abrogate the arationality assumption which is at the heart of contemporary sociological thought. (Something akin to this point was recognized by Mannheim,[18] but his latter-day disciples have tended to assume that one could do sociological history in blissful ignorance of the rational history of ideas!)

We thus see that accepting the arationality assumption has three important consequences: (1) the domain of possible belief situations for sociological analysis is restricted to those in which agents accept beliefs or weight problems in ways incompatible with what rational appraisals would suggest; (2) the sociologist of knowledge must be able to show that the theory of rationality to which he subscribes (in order to determine what cases might be sociological) is the best available one; (3) the historical sociologist of knowledge must show, for any given episode he wishes to explain, that it is incapable of being explained in terms of rational, intellectual history.

In distinguishing between the rational and the socially explicable as I have, I do not mean to suggest that there is nothing social about rationality or nothing rational about social structures. Quite the reverse is the case. The flourishing of rational patterns of choice and belief depends inevitably upon the pre-existence of certain social structures and social norms. (To take an extreme example, rational theory choice would be impossible in a society whose institutions effectively suppressed the open discussion of alternative theories.) Equally, the efficient functioning of most social institutions (e.g., the system of trial by jury) presupposes that the agents within such institutions can, more often than not, make rational decisions.

But this continuous interpenetration of "rational" and "social" factors should not hinder our capacity to invoke the arationality assumption. As John Stuart Mill pointed out more than a century ago, in offering any explanation of some event or belief, *we must not aspire to completeness.* To give a "full" explanation of any situation, *S,* would presumably require a complete enumeration of all the events which have happened in the universe prior to *S,* since all those events are links in the causal chain culminating in *S*. Rather than aspiring to such full explanations, Mill argued that when we explain any situation, *S*, we should select from among the antecedents of *S*, those particular circumstances, *c*, which seem to be most crucial and relevant to the occurrence of *S*. If we take Mill's analysis seriously (and a failure to do so would result in explanatory anarchy) it provides grounds for avoiding that muddle-headed eclecticism which argues that intellectual and social factors can never be usefully distinguished.

Following Mill's lead, we can grant that certain social factors may well be preconditions for rational belief and yet still legitimately exclude those social factors from an explanation of a certain belief, *provided* we can show that the most crucial and relevant antecedent to the acceptance of the belief was a well-founded reasoning process on the part of the believing agent. In thus arguing (as the arationality assumption suggests) for the priority of rational over social explanations for a belief—where both are available—one is not implying that rational decision making has no social dimensions; rather one is stressing that, in those cases where agents have sound reasons for their beliefs, those reasons are the most appropriate items to invoke in an explanation of the beliefs which those reasons warrant.

The Historico-Social Assumption. If the failure to recognize the dependence of cognitive sociology on theories of rationality has been one persistent feature of the sociology of knowledge, another major source of ambiguity is to be found in the tendency to equivocate between, and sometimes to identify, the "historical" with the "social." Karl Mannheim's writings provide ample illustrations of this equivocation. As Mannheim points out, there are two very different sorts of beliefs which we find espoused in the past: those whose very formulation and whose presuppositions are clearly traceable to a particular time and place, and those which betray virtually nothing about their historical or social origins. Putting the distinction a bit differently, one might say that certain propositions inherited from their past carry something of their history *with* them, while others give no clues as to when, and under what circumstances, they were first enunciated. If, for instance, we encounter the statement "the heart is like a pump," we know perfectly well that such a statement must have been made after the invention of pumps, and presumably after some detailed anatomical investigations investigations of the circulatory system. It is simply not a statement which a Greek of the third century B.C., nor a Polynesian of the eighteenth century could have made. At the other extreme, certain beliefs (e.g., "2 + 2 = 4") tell us very little indeed about either the time or the place where they first emerged.

We might call beliefs which do carry their history with them, *contextual,* for they do provide important clues as to the cultural context which generated them. Other beliefs we might call *non-contextual.*[19] Clearly, these two extremes are ideal cases; for the working historian, almost every case will be a matter of greater or lesser contextuality. (Even in the extreme case of beliefs such as "2 + 2 = 4," we can derive reliable conclusions about certain intellectual characteristics of the cultures in which such beliefs could arise.)

What is important here is not the distinction itself, but rather what cognitive sociologists of knowledge seek to do with it. Mannheim, for example, argues that a belief which is contextual (in the sense just sketched above) is a belief which is "historically and socially determined." Given a sufficiently loose sense of "determination," this argument is doubtless sound, indeed vacuously so. But Mannheim's next step is to argue that any contextual belief—that is, any belief which can be definitely located *in history*—is thereby open to *sociological* analysis. If we can pin a belief down to "a particular historical setting," then we have, Mannheim claims, presumptive evidence "of an 'infiltration of the social position' of the investigator into the results of his study."[20]

This argument is entirely specious, precisely because in making it, Mannheim (like others who follow him) trades on an equivocation between the "historical" and the "social." If, for instance, we encounter a statement such as "electricity is caused by a fluid whose particles mutually repel one another," anyone familiar with the history of the physical sciences can readily date it approximately and make some reliable conjectures about the intellectual context in which the statement first arose. Similarly, if we meet a statement like "the Absolute is pure Becoming," anyone familiar with the history of philosophy can readily make sound conjectures about when, where, and by whom that statement became an object of belief. But the fact that these statements are contextual, that they have only been believed at certain times and places, does *not* establish any interesting sense in which they are necessarily *social,* or open to sociological analysis. What makes Mannheim's argument seem initially plausible is his constant conjoining of the terms "historical" and "social," as when he regularly speaks of "historically and socially determined beliefs."[21] He spends great effort quite rightly establishing that certain beliefs are of an *historical* character. It is then by a purely rhetorical slide that Mannheim is able to pretend he has thereby shown such beliefs to be also socially determined in character.

No less a thinker than Emile Durkheim exhibits a similar tendency to assume that any belief which arises in a particular culture or at a particular time must of necessity be socially produced. For instance, in Durkheim's influential *Elementary Forms of the Religious Life,* he claims that certain cultural differences in the laws of logic "prove that they depend upon factors that are historical *and consequently social.*"[22] The last italicized portion of this passage gives the game away. If the establishment of the historical contextuality of a belief is tantamount to rendering that belief socially determined, then the cognitive sociologist has an easy task. He need only look to the history of ideas to find those beliefs which are contextual and he has—hey presto—a whole set of "sociological" desiderata.

As we have said before, however, the slide from the historically to the socially determined is nothing more than intellectual legerdemain. The "consequently social" in the Durkheim passage quoted above is completely gratuitous; if one is to establish that any belief is socially determined, one must—at a minimum—establish some connection between the social situation of the believer and the belief he espouses. The fact that he espoused the belief in 1890 rather than 1870—which is enough to establish the *historical* character of the belief—leaves the question of its social character entirely open.

There are many other cognitive sociologists, besides Mannheim and Durkheim, who seem to believe that if a belief emerged within a particular historical context, then that belief is *a fortiori* susceptible of sociological explanation.[23] But this assumption involves *a confusion of the intellectual culture with the social culture.* As Part One has made clear, it is very often the case that certain beliefs tend to emerge under specific intellectual circumstances which are a function of both the empirical problems recognized in the period and of the dominant research traditions characteristic of the period. But there may be nothing of social or sociological interest about this process of intellectual assimilation of ideas within a pre-established intellectual context or framework.

The Interdisciplinary Assumption. Thus far, we have looked at some of the ambiguities implicit in the historico-social assumption and at some of the difficulties posed by the arationality assumption. There is yet another pervasive assumption about the scope of cognitive sociology which we might call "the interdisciplinary assumption." In its most general form, it assumes that whenever thinkers in one branch of inquiry or discipline draw on, or react to, ideas in other disciplines, then we have grounds for presuming that sociological factors are at work. The more specific version of this postulate, when applied to the history of science, amounts to the claim that *whenever "scientists" are influenced by the "nonscientific"* (e.g., moral, religious, epistemological, metaphysical) *consequences of scientific theory, then this indicates the intrusion of extra-rational, social factors into the scientific situation.*

The interdisciplinary postulate arises, I believe, from an idiosyncratic interpretation of the arationality assumption. If one assumes that science is rational only insofar as it is self-contained, and if one also assumes that whatever is arational is thereby socially caused, then the interdisciplinary assumption follows without difficulty. It is the first premise which makes the inference untrustworthy. As Part One of this work makes clear, it is not necessarily arational for scientists to be concerned about the conceptual relationships between their scientific work (in the narrow sense of that phrase) and the broader intellectual components of contemporary culture. We have already discussed the merits of this claim earlier. What should be pointed out here is that there are whole "schools" of cognitive sociology (one thinks especially of Sorokin, Scheler, Durkheim[24] and Richter, for example) which see the central aim of sociology to be a study of the ways in which the different ideological elements in a culture are integrated. If the arguments of this essay have any cogency, studies of "ideological integration," in so far as such integration is rationally well-founded, belong to intellectual history and fall completely *outside* the realm of cognitive sociology.

It might be thought that these abstract considerations have little bearing on the actual research done by historically oriented sociologists and that these foundational confusions pose no problems when applied to particular cases. Such a view would be quite misleading, as we can see by looking in detail at two of the best-known recent historical studies on the sociology of scientific ideas; namely, the work of Theodore Brown and Paul Forman.

These two historical studies, although concerned with different epochs and different sciences, both seek to show how the reception of certain scientific theories was crucially dependent upon social and institutional circumstances. It is worth analysing these investigations in some detail, for they highlight some of the confused assumptions lying behind even the most sophisticated studies in the historical sociology of science.

Brown's aim is to explain why certain prominent English physicians and natural philosophers enthusiastically accepted the mechanistic approach to life in the middle of the seventeenth century. In brief, his answer is that these thinkers were associated with the Royal College of Physicians, an organization whose social prestige and monopoly over the licensing of medical practitioners were acutely threatened—in part because the College was associated with a moribund and old-fashioned type of Galenic-Aristotelian physiology. The mechanical philosophy, by contrast, was perceived as an up-to-date, "trendy" approach with which the physicians could combat their traditional adversaries—the apothecaries. Brown suggests that the endorsement of the new mechanistic approach to physiology by members of the College was a direct consequence of the institutional and social crisis confronting the College. In Brown's own words: "the collegiate physicians . . . borrowed ideas from the mechanical philosophy . . . because they were engaged in political struggles with their professional prestige seriously lowered and because by borrowing they hoped to raise their prestige again, thereby improving their political position."[25]

Forman, on the other hand, seeks to explain why the indeterminacy principle was so readily and quickly accepted by German theoretical physicists in the late 1920s. Forman's hypothesis is that those physicists were predisposed to be sympathetic to assaults on the causal principle because there was in the German intellectual milieu a strong current (deriving especially from Spengler) which argued that science was overly rationalistic, overly mechanical, overly deterministic—that, in short, it left no room for human values nor for the frailty of the human mind. On Forman's account, this neo-romantic, anti-mechanistic movement threatened the prestige of physical scientists to such a degree that they were actively seeking ways to improve their image by repudiating that deterministic materialism of which they stood accused.[26] The uncertainty relation (when naively interpreted) offered them a splendid riposte to their detractors since the physicists could use it to prove that they were not wedded to a fully mechanistic world picture.

What lies behind the analyses of both Brown and Forman are a set of historiographical assumptions about the character of science, assumptions which allow them to set up their problems in the ways they do. Chief among these assumptions are the Kuhnian convictions that: (1) disciplines generally possess an autonomy which makes them immune from "outside pressures" coming from the broader social and cultural environment;[27] (2) every scientific discipline is fundamentally *conservative,* and resists any re-orientation of its conceptual commitments except in times of acute crisis; (3) these rare periods of intellectual crisis (and here Brown and Forman depart from Kuhn) are generated not from within a discipline, but rather by some external threat to the prestige, funding or intellectual standing of the discipline's practitioners;[28] and (4) the re-ordering of the beliefs of a community of scientists is caused by these outside social pressures rather than any process of rational appraisal within the discipline itself. Forman himself makes many of these presuppositions explicit when he writes:

> We may suppose that *when scientists* and their enterprise *are enjoying high prestige* in their immediate (or, otherwise most important) social environment, *they are also relatively free to ignore the specific doctrines, sympathies, and antipathies which constitute the corresponding intellectual milieu.* With approbation assured, they are free of external pressure, free to follow the internal pressure of the discipline—*which usually means free to hold fast to traditional ideology and conceptual predispositions.* When, however, scientists and their enterprise are experiencing a loss of prestige, they are impelled to take measures to counter that decline . . . [which] may even affect the doctrinal foundations of the discipline . . .[29]

It is worth noting at the outset that neither Forman nor Brown explores whether the emergence of acausal theories in German physics or mechanistic theories in British physiology might have been an entirely appropriate and rational response to the empirical and conceptual criticisms of previously dominant theories. They evidently jump immediately to the assumption that social forces were at work because of their commitment to the thesis that disciplines only allow for the intrusion of nondisciplinary considerations (e.g., of a philosophical, cultural, or political nature) when the discipline in question is under acute social pressure. Equally, their conviction that disciplines are reactionary and resistant to change makes it almost inevitable that, when profound conceptual changes occur within a discipline, they as historians will look to outside social and institutional factors for an explanation of what must seem (on their model of change) to be uncharacteristic, even "unscientific," behavior.[30]

In crucial respects, therefore, the Forman-Brown investigations hang upon the adequacy of their historiographical assumptions (1) to (4). To the extent that the latter are dubious (as I have shown them to be in Part One), the historical researches done under their aegis must remain unconvincing.

Because their (Kuhnian) image of science precludes them from believing that scientists can ever have good scientific reasons for changing their minds, or for worrying about broader intellectual issues, both Forman and Brown systematically ignore the scientific and rational merits of the ideas they discuss. After all, it *might* just be the case that Heisenberg enunciated the indeterminary principle because, as he tells us, he thought that the weight of argument favored it. It *might* just be that Walter Charleton accepted the mechanical philosophy because —as he explains in 400 turgid pages—that theory was rationally preferable to its alternatives. The invocation by Forman and Brown of social and institutional explanations takes place in a curious intellectual vacuum. They do not ask themselves whether their "social" accounts of theory reception do or do not succeed at explaining dimensions of the historical situation which might be explained in terms of sound cognitive reasons. They produce no evidence for their core historical conviction that science is intrinsically conservative and, in usual circumstances, entirely autonomous.[31]

The Theoretical Foundations of Cognitive Sociology

The social causation of ideas. Up to this point, we have been preoccupied with preliminaries, important ones to be sure, but we have yet to say anything about the *content* of sociological theories. If our aim thus far has been to get a little clearer about the problem situations with which a cognitive sociologist ought, in principle, to be concerned, we must now turn our attention toward the character of sociological theory itself. Although this is not the place for anything like a detailed treatment of the substantive commitments of cognitive sociology, a few general observations are perhaps in order, particularly about the cognitive sociology of science.

As we have already noted, any cognitive sociological explanation must, at the very least, assert a causal relationship between some belief, x, of a thinker, y, and y's social situation, z. It will (if the explanations of sociology are in any sense "scientific") do so by invoking a general law which asserts that all (or most) believers in situation type z adopt beliefs of type x.

Hence, the viability of cognitive sociology depends upon our ability to discover general causal (or functional) relationships between social structures and beliefs. More specifically, cognitive sociology of science is predicated on the existence of determinable correlations between the social background of a scientist and the specific beliefs about the physical world which he espouses. Despite decades of research on this issue, *cognitive sociologists have yet to produce a single general law which they are willing to invoke to explain the cognitive fortunes of any scientific theory, from any past period.* The acceptance of Boyle's law, the rejection of Lamarck's theory of heredity, the reception of Lyell's geology, the genesis of Newton's ideas, the repudiation of Galenic physiology, the historical career of the theory of relativity—these are but a tiny sample of the cases where contemporary sociological theory has failed to provide any historically significant aids to the understanding. When sociological explanations of specific cases are offered, the reader is generally left to guess for himself what principles they presuppose.[32]

Nor should one be surprised at the exegetical bankruptcy of contemporary, cognitive sociology of science, for its current explanatory repertoire is far too crude to permit the kinds of discriminations that are called for. Whether we talk about social classes, economic backgrounds, systems of kinship, occupational roles, psychological types or patterns of ethnic affiliation, we find that these generally bear no close relation to the belief systems of major scientists. Sons of working class men as well as of noblemen are found among both defenders and detractors of the Newtonian theory in the eighteenth century; politically conservative as well as politically radical scientists accept Darwinism in the 1870s and 1880s. Followers of Copernican astronomy in the seventeenth century represent the entire spectrum of occupational roles from university don (Galileo) to gentlemen-soldier (Descartes) to priest (Mersenne), and of psychological types.

A judicious examination of the historical record seemingly undercuts the effort to link major scientific theories to any particular socio-economic group. The Marxists are simply wrong in speaking of a specifically bourgeois mathematics; the followers of Weber have presented no convincing evidence for the existence of a specifically Puritan natural philosophy; contrary to fascist ideology, there is no distinctly Jewish physics; against the claims of many Leninists, we have no evidence that there is a specifically proletarian version of the special theory of relativity.

The chief reason for the sociologists' failure to find a correlation between scientific belief and social class is that the vast majority of scientific beliefs (though by no means all) seem to be of no social significance whatever. That gravity obeys an inverse square law, that mechanical energy can be converted into heat, that atoms have nuclei; such beliefs seem (and I stress the verb) to have no conceivable social roots or social consequences. Given the evident conceptual distance between most scientific beliefs and the vagaries of social change, it is very different to imagine how social pressures can have been responsible for the generation or reception of such ideas. To make matters worse, contemporary sociology does little to clarify, even in the abstract, the mechanisms whereby social factors might influence the adoption of specific scientific ideas. Whether we look to Marx, Mannheim, Merton or to any of the other leading sociological theorists, we are left completely in the dark when it comes to the specification of a general mechanism for explaining the connection between social situation and ideological commitment in the scientific or philosophical sphere. Why should (to take some standard examples) living in a mercantile society incline one to favor empiricism? Why should living in a feudal society dispose one to a geocentric theory of the universe? Why should—to use a notorious example from Hessen—the fact that Newton lived in a seafaring nation cause him to interpret Boyle's law in the way he did?[33] What evidence we do have suggests that patterns of scientific belief, both rational and arational, cut across all the usual categories of sociological analysis. It is presumably for just such reasons as these that many contemporary sociologists of science (such as Ben-David and, in certain moods, even Merton and Mannheim) hold out little hope for the cognitive sociology of science. As Ben-David puts it, "The possibilities for . . . [a] sociology of the conceptual and theoretical contents of science are extremely limited."[34]

Confronted by the widely acknowledged failure of contemporary cognitive sociology to explain any interesting scientific episodes, we could draw one of two conclusions:

(a) we might conclude that the failure of cognitive sociology of science is due to the fact that belief-determination in the natural sciences is *instrinsically* immune to sociological influences, and thus to sociological analysis.

Alternatively, we might more charitably suggest that,

(b) there is no reason *in principle* why arational scientific beliefs cannot be explained sociologically, provided we can develop more subtle theories than we presently have about the social causation of scientific belief. Many leading sociologists of science argue for (a), seeing the role of sociology as entirely non-cognitive, at least so far as the natural sciences are concerned.

Robert Merton, for instance, in his classic *Science, Technology and Society in 17th-Century England,* specifically disclaims any ambition to explain the *content* of seventeenth-century science in sociological terms, remarking that "specific discoveries and inventions belong to the internal history of science and are largely independent of factors other than the purely scientific."[35] Karl Mannheim goes so far as to conclude that historical developments in "mathematics and natural science" are "determined to a large extent by immanent factors."[36] Their arguments for this view are unconvincing, however, because they rest on that same naive empiricist conception of science and of scientific rationality which we have discussed before. On the whole, those cognitive sociologists who exclude science from their domain do so because of two related convictions, both of which are seriously wrongheaded:

1. a conviction that scientific theories are dictated by the data, leaving no room for subjective, nonfactual determinants of knowledge; as Maurice Richter puts it, "society cannot, in principle, determine the contents of scientific knowledge, because these are to be determined by observations of nature."[37]

2. a belief that proper scientific knowledge is self-contained and insulated from other strands of human belief (e.g., religion, philosophy, values) which are, in part, socially determined.

It is the conjunction of beliefs (1) and (2) which leads many thinkers to deny the possibility of a cognitive sociology of science. To the extent that both images of science are wrong, as I claim them to be, then there is little warrant for asserting (a) above. Since it has been established that science interacts with other disciplines, then, if we could establish that beliefs in those disciplines were "existentially" determined, it would follow as a matter of course that science too, at least to the extent of its interaction, is (at least indirectly) socially determined. But even if the denial of (1) and (2) allows for the possibility of a cognitive sociology of arational scientific beliefs (i.e., (b) above), it must be stressed that far more *theoretical* work within sociology itself is required before we can get any mileage from cognitive socio-history.

If it is true that many sociologists are pessimistic about the prospects for a cognitive sociology of science, they are generally far more sanguine about a cognitive sociology of such disciplines as theology and philosophy. Unfortunately, their record in these other areas is almost as discouraging as it is in science itself. For instance, in his provocative discussion of the history of epistemology, Mannheim observes, with much justice, that theories of knowledge in the seventeenth century were powerfully influenced by the newly emerging scientific theories of the period. Generalizing this result, he claims that, "every theory of knowledge is itself influenced by the form which science takes at the time and from which alone it can obtain its conception of the nature of knowledge."[38] Mannheim then immediately asserts that this dependence of epistemology on science proves that theories of knowledge are "socially" determined.[39] The only way to make Mannheim's inference even seem cogent is by assuming that it is not immanent or rational for epistemologies to reflect shifts in scientific belief. But if we adopt an alternative model of rationality, we can see that it is often entirely reasonable and natural for a symbiotic interconnection to exist between science and philosophy. The existence of such an interdependence itself entails nothing about whether it is socially caused.

It was argued in the first part of this chapter that the application of sociological analyses to the history of scientific ideas must await the prior development of a rational or intellectual history of science; it should be equally clear that the emergence of a cognitive sociology of knowledge in general history must also await the articulation of some radically new tools and concepts of sociological analysis.[40] Until both these logically prior tasks are well under way, pious assertions about the social determination of scientific belief remain merely gratuitous articles of faith.

Conclusion

Through much of this chapter, I have been highly critical of much work, both theoretical and applied, in the sociology of knowledge. It is crucial to stress, however, that these are objections to the subject as it is normally practiced. Nothing I have said here raises doubts about the possibility of the sociology of knowledge (provided that it works within the framework of the arationality assumption). To the contrary, large scope is given in my account for cognitive sociological research. Whenever, for instance, a scientist *accepts* a research tradition which is less adequate than a rival, whenever a scientist *pursues* a theory which is non-progressive, whenever a scientist gives greater or lesser *weight* to a problem or an anomaly than it cognitively deserves, whenever a scientist chooses between two equally adequate or equally progressive research traditions; in all these cases, we must look to the sociologist (or the psychologist) for understanding, since there is no possibility of a rational account of the action in question. We stand badly in need of sociological theories that can illuminate such cases, which are undoubtedly frequent in the history of thought. Particularly promising here would be an exploration of the *social determinants of problem weighting,* since that phenomenon—probably more than the others—seems intuitively to be subject to the pressures of class, nationality, finance and other social influences.

Equally, we need further exploration *into the kinds of social structures which make it possible for science to function rationally* (when it does so). Although no social system is sufficient to guarantee progress and rational scientific choice, certain sociopolitical institutions are presumably more conducive to achieving those ends than others. Once again, however, we must understand what scientific rationality is before we can study its social background.

Epilogue:
Beyond *Veritas* and *Praxis*

Of the many questions which this excursion has left unanswered, at least two require further discussion:

1. even if we grant that the aim of science is problem solving, and if we grant further that science has been effective at such problem solving, we are entitled to ask whether an inquiry system like science—with the techniques it has at at its disposal—is the most effective possible mechanism for the solution of problems:
2. we are also entitled to ask whether the investigation of intellectual problems of the type science studies can be justified, given the other compelling demands on our limited mental, physical, and financial resources.

Definitive answers to these questions are not readily within reach, but one can at least sketch in what directions we need to move to answer them.

Much has been written of the methods of science, yet with the notable exception of pragmatists like Peirce and certain recent "systems analysts," no one has seriously investigated whether the methods utilized by science are the most conducive to generating solutions to problems. The preoccupation of classical philosophers of science has been with showing that the methods of science are efficient instruments for producing truth, high probability, or ever closer approximations to the truth. In this enterprise, they have failed dismally. What we now need to ask is whether the methods of science—even if they fail as good "truth machines"—are the best available tools for the solution of problems.

That science has solved problems is undoubted; the question is whether any emendations of the traditional tools of empirical and logical appraisal would be likely to increase the problem-solving efficiency of science.

This is not the place to propose answers to these global queries. But we are entitled to urge that the questions themselves are serious ones which should no longer be ignored. Until and unless we can show why science can be an effective instrument for the solution of problems, then its past success at problem solving can always be viewed as an accidental piece of good fortune which may, at any time, simply dry up.

But this, in turn, raises the still larger question we mentioned above: even if science could be shown to be the best tool for the solution of cognitive problems, how can we justify devoting such ample resources to the satisfaction of a peculiar feature of animal evolution, namely, man's sense of curiosity?

Classically, the justification for scientific research was twofold. It was stressed, on the one hand, that man's quest for truth about the world ("knowledge for its own sake") was the driving force behind scientific inquiry. On the other hand, it was urged that science has enormous practical, utilitarian value in improving the physical conditions of life. Both these approaches to the matter have worn thin. Science does not, so far as we know, produce theories which are true or even highly probable. Equally, it is time to acknowledge publicly that the optimistic Baconian identification of knowledge with power is as ill-founded in our time as it was when the Lord Chancellor of England first promoted it some 350 years ago. Much theoretical activity in the sciences, and *most of the best of it,* is not directed at the solution of practical or socially redeeming problems. Even in those cases where deep-level theorizing has eventually had practical spin-off, this has been largely accidental; such fortuitous applications have been neither the motivation for the research nor the general rule. Were we to take seriously the utilitarian approach to science, then a vast re-ordering of priorities would have to follow, since the present allocation of talent and resources within science manifestly does not reflect likely practical priorities.

If a sound justification for most scientific activity is going to be found, it will eventually come perhaps from a recognition that man's sense of curiosity about the world and himself is every bit as compelling as his need for clothing and food. Everything we know about cultural anthropology points to the ubiquity, even among "primitive" cultures barely surviving at subsistence levels, of elaborate doctrines about how and why the universe works. The universality of this phenomenon suggests that making sense of the world and one's place in that world has roots deep within the human psyche. By recognizing that solving an intellectual problem is every bit as fundamental a requirement of life as food and drink, we can drop the dangerous pretense that science is legitimate only in so far as it contributes to our material well-being or to our store of perennial truths. Viewed in this light, the repudiation of theoretical scientific inquiry is tantamount to a denial of what may be our most characteristically human trait.

To say as much is not to suggest that the expenditure of resources on all theoretical problems in science is justified *tout court.* Far too much scientific research today is devoted to problems which are as cognitively trivial as they are socially irrelevant. If the "pure" scientist is to deserve the generous support presently being lavished on him, he must be able to show that his problems are genuinely significant ones and that his program of research is sufficiently progressive to be worth gambling our precious and limited resources on it.

Notes

Prologue

1. Rudolf Carnap, for instance, readily concedes his system of inductive logic and his confirmation theory to be *totally inadequate* for dealing with the more important episodes in the history of science: "For instance, we cannot expect to apply inductive logic to Einstein's general theory of relativity, to find a numerical value for the degree of confirmation of this theory. . . . The same holds for the other steps in the revolutionary transformation of modern physics . . . *an application of inductive logic in these cases is out of the question*" (my italics; [1962], p. 243). Most other proponents of inductive theories of rationality have made similar disclaimers about their models.

2. Carnap, again, finds himself being forced to the view that the degree of confirmation (Carnap's basic measure for rational acceptability) of all universal scientific theories is zero, which is precisely the confirmation they would have if they had never been confirmed at all! Carnap, in a classic piece of understatement, agrees that this "result may seem surprising; it seems not in accord with the fact that scientists often say of a law that it is " 'well-confirmed' . . .": *(Ibid.*, p. 571.)

3. Whether these episodes are genuinely irrational or whether they only seem to be is an issue to which I shall return in chapter seven.

4. See especially Kuhn (1962), and Feyerabend (1975).

5. For a detailed discussion of Kuhn's views on this question, see below, pp. 148ff.

6. Cf. Lakatos (1968b), where he struggles valiantly to make the Popperian theory of rationality germane, and to fit his own interesting ideas into a Popperian context (where they do not really belong).

7. Although Hintikka has avoided some of the difficulties which Carnap encountered, he, like Carnap, retains the view that degrees of confirmation are generally language dependent. This failing is as troublesome and as counter-intuitive as any of Carnap's earlier results.

Chapter 1

1. The two apparent exceptions to this claim are Kuhn and Popper, who both insist that their models of science are based on a problem solving approach to scientific growth. Unfortunately, such overtures to problems are only rhetorical. Popper never convincingly shows how the logic of problem solving relates to any of the technical elements of his philosophy of science (such as "falsifiability" or "empirical content"); Kuhn, for his part, denies that "the ability to solve problems is either the unique or an unequivocal basis for paradigm [i.e., theory] choice" (Kuhn [1962], p. 168). Both thus take away with one hand what they give with the other.

2. This is not to claim, of course, that philosophers of science have ignored the fact that science is empirical. But, as we shall see below, there are vast differences between "explaining empirical data" and "solving empirical problems." Philosophers of science have said too much about the former and virtually nothing about the latter.

3. Cf. Oresme (1968), p. 244. (I am grateful to Dr. A. G. Molland of the E. R. Institute for this reference.) A fascinating account of some of the "nonfactual" phenomena which have been treated as empirical problems by scientists is in Martin (1880).

4. There are other important technical differences between empirical problems and facts (such as that a theory always explains an infinite number of factual propositions but only solves a finite number of problems) which will be discussed later.

5. My category of unsolved empirical problems corresponds approximately to Kuhn's notion of "puzzle." It is important to stress that Kuhn's puzzle solving view of science embraces *nothing other* than this class of unsolved problems.

6. It should be emphasized that this conception of an anomaly is significantly different from the conventional one. (See the next sections for a full discussion of the details.)

7. Once solved by any theory, however, they generally remain as problems which subsequent theories are expected to solve (at least until they can convincingly be shown to be pseudo-problems).

8. A propos the problem of Brownian motion, John Conybeare—a contemporary of Brown's—wrote: "I don't believe a word on't . . . [Biot] states it to be possible that solid bodies may be compared of [*sic*] systems of moving molecules, representing in small what the planetary systems do in large. I would only add one supposition more; that these molecules are inhabited, and have philosophers among their population who . . . believe they have developed the system of the universe." This quotation is taken from Mary Jo Nye's excellent history of the reception of Brownian motion (1972), pp. 21-22. For further discussions of this episode, see Brush (1968).

9. See Vartanian (1957).

10. It is worth pointing out that Lakatos' theory of "research programmes" (for all its stress on competition between theories) cannot explain cases such as

these because materialistic biology did not predict the polyp *in advance of its discovery,* and thus (on his view) can take no credit for being able to explain it.

11. See chapter four below, especially pp. 125-27.
12. Cf. especially Duhem (1954), Neurath (1935), and Quine (1953).
13. Especially Kuhn and Lakatos.
14. Popper has come close to grasping this point, (b′), with his requirement that any acceptable new theory must be able to explain *everything* which its predecessors and competitors can. Unfortunately, however, Popper goes too far, because in his adherence to (a), he makes *any* loss in explanatory content a fatal blow to any theory which exhibits it. By contrast, I am claiming that the loss of explanatory content by virtue of a non-refuting anomaly counts against a theory, but not necessarily decisively. For a fuller criticism of Popper's (and Lakatos') cumulative theory of science, see below, pp. 147-50 and Laudan (1976b).
15. It is important to stress the converse of this point: if a problem has not been previously solved by any predecessor of a theory, then it simply constitutes an unsolved, not an anomalous, problem for that theory (with the proviso that at some later point the problem may cease to be a problem altogether; in which case, of course, it would no longer be anomalous).
16. Indeed, it would probably not be far wrong to identify the historical emergence of a science from a proto-scientific state as that point at which all its problems cease to be of the same weight.
17. Home (1972-73) shows convincingly that Franklin's treatment of the Leyden jar effectively diverted attention away from what had previously been regarded as the central problems of electrical theory. (Cf. especially *ibid.*, pp. 150-51.)
18. Cf. especially Kuhn (1962).
19. See Duhem (1954) and L. Laudan (1965).
20. Many of these claims have been disputed by Grünbaum; see especially (1960), (1969), and (1973).
21. The only way in which some T_i which was a member of the complex C can remove a from among its class of anomalous instances is by the development of an alternative complex C', including T_i, which can turn the anomalous a into a solved problem.
22. And in showing when it is rational to preserve the entire complex and ignore the anomaly.

Chapter 2

1. For a criticism of Kuhn's views on this matter, see below pp. 150-51, 173-75.
2. Karl Popper, for instance, has often insisted that the use of metaphysical or theological beliefs to criticize scientific theories is only of "sociological" interest and is in no way germane to the understanding of rational evaluation. In one of his most recent essays, for example, Popper writes: "the historical and sociological fact that the theories of both Copernicus and Darwin clashed with religion is completely irrelevant for the rational evaluation of the scientific theories proposed by them" ([1975], p. 88). In a slightly different vein, Philip

Frank—confronted by the failure of Renaissance astronomers to accept Copernicanism—argues that they made the choice by asking "whether the life of man would become happier or unhappier by the acceptance of the Copernican system" ([1961], p. 17). Frank allows for no middle ground between a purely "scientific" (i.e., empirical) evaluation, on the one hand, and hedonistic value judgments, on the other.

3. The most interesting recent exception is Gerd Buchdahl who has discussed at length (see especially [1970]) the role of controversies about nonempirical issues in the history of science. My account of conceptual problems, although different from Buchdahl's, owes a great deal to his sensitive treatment of these issues.

4. Heimann (1969-70) utilizes the search for internal consistency as a means of explaining the evolution of Maxwell's views on electricity and magnetism.

5. It should be noted, however, that the refusal to *accept* an inconsistent theory need not require that one cease working on such a theory. (See below, pp. 180ff.) On the role of internal conceptual problems in the development of Thomas Young's work, see Cantor (1970–71).

6. See Hare (1840), and Faraday's perplexed response (1840) to this conceptual criticism.

7. See especially Stallo (1960).

8. See Whewell (1840), part II. For an excellent account of Whewell's analysis see Butts, forthcoming.

9. The most common form of mutual reinforcement between theories is that relation usually known as "*analogy.*" (For an interesting demonstration of how crucial this sort of analogical problem was in nineteenth century chemistry, see Brooke [1970–71.)

10. Viner (1928) offers a convincing argument that one of the central conceptual problems for Adam Smith's economic theory was its incompatibility with the Newtonian thesis of a balance of forces in nature. The issue was particularly acute since Smith's economic theory relied on a general (Newtonian) balance of nature and yet postulated forces of economic motivation (e.g., self-interest) which were seemingly incompatible with such a balanced system. It has been argued that Smith wrote his treatise on moral philosophy in order to resolve this tension.

11. For instance, any astronomical claim based on telescopic observation presupposes the acceptability of certain optical theories. The best general discussion of the conceptual and experimental interdependence of the physical sciences is still Duhem (1954).

12. Koyré put the point this way, "abstract methodology has relatively little to do with the concrete development of scientific thought" ([1956], p. 13).

13. To mention only a few examples: Buchdahl (1969) and Sabra (1967) have examined the role of methodology in seventeenth century mechanistic science; Cantor (1971), Olson (1975) and Laudan (1970) have studied the impact of the epistemology of the Scottish school on reception of physical theories in the late eighteenth century; McEvoy and McGuire (1975) have

explored the relations between Priestley's methodology and phlogistic chemistry; Brooke (1970–71) has analyzed the impact of Comtean positivism on nineteenth century French chemistry and physics; Hooykaas (1963) and R. Laudan (forthcoming) have studied the impact of methodology on geology in the Lyellian period; Buchdahl (1959), Knight (1970) and L. Laudan (1976a) have analyzed the methodology of the atomic debates; Hull (1973), Ellegard (1957), Ghiselin (1969) and Hodge, (forthcoming), have documented the impact of methodological ideas on Darwin and his critics.

14. See Cantor (1971) and L. Laudan (1970).
15. See L. Laudan (1973b) and (1977).
16. Buchdahl (1970).
17. McGuire and Heimann (1971).
18. See especially Cotes' preface to the second edition of Newton's *Principia*.
19. This point is cogently argued in McGuire and Heimann (1971).
20. For a brilliant study of the role of epistemological and metaphysical issues in eighteenth century embryology, see Roger (1963). Roger's treatment of Buffon provides an ideal working model of the type of conceptual historical analysis for which this chapter seeks to provide a rationale.
21. A contemporaneous example of world-view difficulties can be found in Culotta's suggestive study (1974) of nineteenth century biophysics.
22. Some members of this group flatly deny that the evolution of science owes anything to the broader backdrop of philosophical convictions; others (such as Duhem) recognize the impact of philosophy on science, but bemoan it.
23. See above pp. 36–40.

Chapter 3

1. Cf. especially Shapere's excellent critique (1964), and Masterman (1970). The ambiguity of Kuhn's analysis has been multiplied as a result of Kuhn's later retractions of many of the basic ideas of the first edition of his *Structure of Scientific Revolutions* (1962). Unable to follow the logic of his later changes of mind, I have been forced to characterize Kuhn's views in their original form.
2. For a criticism of Kuhn's theory of "mature" science, see below pp. 150–51.
3. It should be stressed that Kuhn's notion of "anomaly" is the traditional one (anomaly=refuting instance) rather than the one I have sketched above, pp. 26ff.
4. "If any and every failure to fit [the facts] were ground for theory rejection, all theories ought to be rejected at all times" (Kuhn [1962], p. 145).
5. As Kuhn originally put it: "there are losses as well as gains in scientific revolutions" ([1962], p. 66). Kuhn, however, is not altogether consistent on this issue (see below p. 237 n. 18).
6. Shapere (1964).
7. See especially Feyerabend (1970c).
8. Cf. the post-script to the second edition of Kuhn's (1962).

9. Kuhn (1962), p. 42.
10. Cf. Lakatos (1970), pp. 133-34.
11. *Ibid.*, p. 135.
12. *Ibid.*, p. 118.
13. See below, pp. 140ff.
14. In spite of the universally acknowledged—and *prima facie* insoluble—difficulties facing anyone who would make comparisons of the logical or empirical content of actual scientific theories, virtually all the recent discussions of scientific growth to have come out of the Popperian tradition—including those of Popper himself, Watkins, Lakatos, Musgrave, Zahar and Koertge—still assume that the touchstone for scientific progress is increasing content.
15. Cf. especially Grünbaum (1976a).
16. Despite much special pleading and handwaving, neither Lakatos' study of Bohr (1970), nor Zahar's study of Lorentz (1973), nor the Lakatos-Zahar study of Copernicus (1975) utilizes Lakatos' "official" theory of progress. At *no* point do they show those relations of content-inclusion which are crucial to progress (in Lakatos' sense).
17. Nor can Lakatos use those assessments for explaining the actions of scientists since he denies that anything except *retrospective* autopsies of long-dead scientific controversies can produce a reliable assessment.
18. I may be unfair to Lakatos here, for he wildly equivocates on this issue. On the one hand, he insists that the unfalsifiable hard core of a theory is one of the central features of a research programme from its inception. On the other hand, he tells us that "the actual hard core of a programme does not actually emerge fully armed . . . [it] develops slowly" ([1970], p. 133n). If, in fact, the hard core cannot be identified during much of the history of a research programme, then how do scientists know what to hold dear when confronted by an anomaly?
19. In her study of eighteenth century mechanics, Iltis (1972-73) seems to find it strange that scientists who accepted Newton's or Leibniz's mechanics also tended to accept the ontology, the methodology, and even the theology, associated with these theories. The doctrine of research traditions makes this surprising phenomenon completely natural and unsurprising.
20. For a useful account of seventeenth-century optics, see Sabra (1967).
21. McKie and Partington (1937-39).
22. For a discussion of what is involved in abandoning truth and falsity as determinable characteristics of research traditions, see below pp. 124ff.
23. Historians who focus on specific theories, rather than the larger traditions of which they are a part, often find themselves perplexed by, and unable to explain, the reception of such theories. Such perplexity often dissolves if these theories are seen in a larger context. For instance, Alan Shapiro's excellent study of seventeenth-century wave optics (1973) ends with a "paradox"; Huygens' theory of light, as Shapiro rightly argues, was the only theory available at the time which could explain the double refraction in Iceland spar. Why, then, asks Shapiro, was Huygens' approach so totally

ignored for the next century and why did scientists remain committed to a Newtonian approach (which could not do justice to the problems posed by double refraction)? Shapiro offers no answer. Surely a part of the answer is to be found in the fact that Huygens' theory—although it could (even this was sometimes doubted) handle Iceland spar—was found wanting because it refused to address itself to, or to offer any solutions for, most of the important problems in late seventeenth-century optics. (For instance, it did nothing to solve the problems of colors or Newton's rings.) Equally, it was thought to suffer from some serious anomalies (e.g., its inability to explain the sharp lines around shadows). If we add to this the fact that Huygens' optical work was allied to a broader Cartesian tradition in optics—a tradition far less progressive than Newton's—it is not surprising that Huygen's *Traité de la lumière* "rapidly sank into oblivion" (Shapiro [1973], p. 252). One could even go so far as to say that Huygens' theory was not taken seriously because, given the flaws mentioned above, *it did not deserve to be taken seriously.*

24. Cf. Brown (1968).

25. As we have already indicated, the methodology of the uniformitarian tradition in geology (as developed by Hutton, Playfair, and Lyell) decreed that *all* the problems of cosmogony—which had previously been regarded as geological problems—would no longer have to be solved by geologists.

26. For an interesting account of the fortunes of aether theories in the late nineteenth century, see Schaffner (1972). For a discussion of empirical problems which "vanish," see Grünbaum (1976a).

27. See especially Cantor (1971).

28. Cf. L. Laudan (1970), (1973b), and (1977).

29. Lakatos has been misled by this feature of research traditions into thinking that empirical anomalies are of virtually *no significance* to the development of science. Quite the reverse is the case, for at least two reasons:

a. It sometimes happens that the heuristic capacity of a research tradition is too weak to allow it to accommodate certain anomalies, and its failure to deal with them convincingly weighs heavily against it.

b. Even when a research tradition is sufficiently fecund to provide guidelines for transforming some anomalous problem into a solved one, the existence of the anomaly is historically crucial if we wish to understand why the theories within a research tradition exhibit the *sequential* character they do. Contrary to Lakatos' a priorism, the order of theories constituting a research tradition will mirror, at least partially, the order in which different anomalies emerged.

30. There are unmistakable ambiguities in Lakatos' treatment of this question. On the one hand, Lakatos characterizes a research programme chiefly in terms of its so-called hard core, i.e., those doctrines which are so crucial to the programme that no scientist within the programme will abandon them. On the other hand, Lakatos insists that "the actual hard core does not actually emerge fully armed . . . it develops slowly, by a long, preliminary process of trial and error" ([1970], p. 133n). This latter approach suggests that research programmes have *no* "hard core" in their early stages; but if that is true, then

how can Lakatos identify research programmes in their infancy, since that identification depends upon specification of the contents of the hard core? (Cf. note 18 above.)

31. For an illuminating analysis of the manner in which the core assumptions of a research tradition can undergo radical transformation, see Brown's study (1969) of theories of the electric current in the early nineteenth century.

32. As Hull has cogently argued "no degree of similarity between earlier and later stages" in the development of an historical "object" such as a research tradition is necessary in order for it "to remain the same entity" ([1975], p. 256).

33. Despite Lakatos' contempt for the method of trial and error, his only explanation for the emergence of the core of a research tradition is that it results from "a long, preliminary process of trial and error" ([1970], p. 133n).

34. Indeed, if Forman (1971) is right, the abandonment of strict determinism in modern quantum mechanics was prompted by the irreconcilability of classical physics with the general world-view.

35. (1961), p. 191.

36. Schofield (1970).

37. See above, pp. 66-69.

38. My analysis here owes much to discussions with Adolf Grünbaum.

39. I find it very difficult to pin down precisely what Kuhn's views on this issue actually are. Consider, for instance, the following remark: "Though the historian can always find men—Priestley for instance—who were unreasonable to resist [a new paradigm] for as long as they did, he will not find a point at which resistance becomes illogical or unscientific." (Kuhn [1962], p. 158). The first half of the passage suggests that there are criteria for determining whether the acceptance or rejection of a paradigm is rational; whereas the final clause denies that there is any point at which that acceptance becomes rational (assuming, as I think we are entitled to, that Kuhn is here using "unreasonable," "illogical," and "unscientific" as approximate synonyms). But if there is no point at which the acceptance (or rejection) of paradigm becomes reasonable, how could we decide—as Kuhn has—that Priestley was "unreasonable" in rejecting Lavoisier's paradigm?

40. Like Feyerabend, Kuhn recognizes that there is a context of pursuit and denies that there are usually any rational grounds for pursuing a new theory which has not yet been well confirmed: "the man who embraces a new paradigm at an early stage must often do so in defiance of the evidence provided by problem-solving [success] . . . *A decision of that kind can only be made on faith*" (Kuhn [1962], p. 157; my italics).

41. In a famous paper published in 1813, the Swedish chemist Berzelius discussed many of the anomalies for Daltonian atomism. However, precisely because "it would be rash to conclude that we [atomists] shall not be able hereafter to explain these apparent anomalies in a satisfactory manner" ([1813], p. 450), Berzelius did not urge non-pursuit of the atomic theory even though, within the context of acceptance, "the hypothesis of atoms can neither be adopted nor considered as true" *(Ibid.)* Cf. also Berzelius (1815).

42. See, for instance, A. Grünbaum (1973), pp. 715–25, 837–39; I. Lakatos (1970); E. Zahar (1973), especially 100ff.; K. Schaffner (1974), especially 78–79; and J. Leplin (1975). A thorough historical analysis of the evolution of the notion of adhocness would probably show that the idea originated at a time when scientists and philosophers believed: (1) that the constituent parts of a theory could be tested in isolation; and (2) that only *directly observable* entities could be legitimately postulated within a theory. Most philosophers and scientists have now abandoned both (1) and (2), yet continue to believe that the requirement of independent testability is still legitimate. Whether the continued demand for the latter makes any sense given the repudiation of the simple-minded philosophy of science which originally motivated it is an open question. [Grünbaum's 1976b appeared too late for me to discuss it here. —Au.]

43. See the writings of Lakatos and Grünbaum cited above, as well as the relevant sections of Karl Popper (1959) and (1963).

44. Cf. Grünbaum (1973), p. 718. (Although this useful clarification is due to Grünbaum, it does not represent his own approach to the problem.)

45. Cf. above pp. 40–44.

46. A fuller treatment of this problem is in L. Laudan (1976b).

47. Such a context—and comparison—dependent sense of ad hoc is discussed sympathetically in Grünbaum (1973).

48. Utilizing the machinery outlined above, pp. 66–69.

49. Zahar, for instance, speaks of a theory being ad hoc "if it is obtained from its predecessor through a modification of the auxiliary hypotheses which does not *accord with the spirit* of the heuristic of the [research] programme" ([1973], p. 101; my italics). On another occasion, he suggests that a theory is ad hoc in this sense if it "destroys the organic unity of the whole nexus" *(Ibid.,* p. 105). Zahar may have clear criteria for these processes, but he never unpacks what it would mean to be out of "accord" with "the spirit of a programme's heuristic" or to destroy its "organic unity." Schaffner is slightly more specific, suggesting that theories can encounter "trans-empirical" difficulties such as "complexity" or "theoretical discord"; but until these notions are further developed one cannot be sure whether Schaffner has in mind the same sort of analysis for which I have argued here.

Chapter 4

1. For a discussion of some of the weaknesses in classical theories of self-correction and truth-approximation, see L. Laudan (1973a). A devastating critique of Popper's theory of verisimilitude is Grünbaum (1976c).

2. Maxwell has attempted to defend the view that it is rational to seek a goal (such as truth) "even though we have no rational assurance whatsoever that the aim will meet with success" ([1972], p. 151). It is just such an argument as this which lies behind beliefs in immortality, the philosopher's stone and El Dorado. It argues that quixotic quests are always rational until we can prove them to be otherwise. Surely the burden of proof is precisely reversed; hunting

the snark does not become rational just because we have not yet proved its nonexistence!

3. Scheffler (1967), pp. 9–10.

4. Evidently in fear and trembling lest the incorporation of these evolving standards into a model of rationality might deprive it of its supra-temporal ("third-world") status, they have deliberately repudiated the use of such notions, taking refuge, rather, in what they imagine to be non-"time-dependent propert[ies]" (Zahar [1973], p. 242n.; see also Lakatos [1970], p. 137) such as "mathematical coherence." Leaving aside the dubious contention that conceptions of mathematical coherence themselves have not evolved, one wonders what point there is in maintaining that all the significant meta-level characterizations of science have been static since the garden of Eden.

5. This model thus allows us to have best of both worlds; we can acknowledge that specific standards of rationality have evolved, without surrendering our capacity to make normative judgments about the past. It is not uncommon to find in the sociological literature a distinction (similar to the one I have sketched) between rationality *within* a given context of belief and what is frequently called "transcendent rationality" (See, for instance, Winch [1964] and Lukes [1967].) What has not been suggested before, so far as I can determine, is that there is a third, hybrid sense of rationality which allows us to make transcendent judgments about the rationality of beliefs without ignoring the crucial particularities of context.

6. "It is precisely the abandonment of critical discourse that marks the transition to a science . . . [thereafter] critical discourse recurs only at moments of crisis when the bases of the field are again in jeopardy" (Kuhn [1970], pp. 6–7).

7. Truesdell (1968), the well-known historian of eighteenth-century mechanics, does his best to play down many of these issues, particularly those which are not mathematical in character. Costabel (1973) and Aiton (1972) give much more sensitive accounts of some of the philosophical issues at stake in the mechanics of the Enlightenment.

8. Concerning ontology, cf. especially McGuire and Heimann (1971) and Schofield (1970). On methodology, L. Laudan (1973b) and (1976). See also above, pp. 57-61.

9. Cf. Kuhn (1962), p. 10.

10. Kuhn's cynical view is that scientific revolutions are regarded as progressive because the "victors" write the history and they would hardly view their own successes as anything but progressive. (Cf. especially his [1962], pp. 159ff.) Here, as elsewhere, Kuhn slides too readily between political and cognitive characterizations of science.

11. See below, pp. 147–50.

12. There are excellent summary discussions of the difficulties of the implicit definition theory of meaning in Suppe (1974), pp. 199ff, and in Shapere (1966).

13. Kuhn (1970), p. 266.

14. If the theoretical assumptions are inconsistent with the theory under analysis, then the problem will become a "pseudo-problem."

15. In making these determinations, we would have to restrict ourselves, of

course, to those problems and anomalies capable of being expressed within the framework of the research tradition under scrutiny and would have to ignore rival and (by hypothesis) incommensurable research traditions. The possibility of assessing these variables does depend upon translations being possible between the theories that constitute a research tradition.

16. My approach to the problem of incommensurability resembles that of Kordig (1971), insofar as we both argue that there are *methodological* criteria for theory comparison, even when substantive translation between different theories is inappropriate. Kordig and I differ drastically, however, about what these methodological criteria should be. Following Margenau, Kordig stresses a comparison of theories with respect to their empirical confirmation, their "extensibility," their "multiple connection," their simplicity and their "causality"; unfortunately, most of these remain entirely intuitive notions in Kordig's discussion, and it is to be hoped that he will refine them into the sensitive instruments of analysis needed for the comparative appraisal of theories.

17. The cogency of this argument does *not* rest on the acceptance of the model outlined in this essay. *Any* model of rationality which offers a method of determining an appraisal measure of scientific theories without inter-theoretic translation can avoid the difficulties of incommensurability.

18. Here, as elsewhere, Kuhn is ambivalent. On the one hand, he stresses the non-cumulative character of science by insisting that there are always problem losses as well as gains in every case of paradigm replacement. (See above, p. 231 n. 5.) Yet, on the other hand, he claims that: "A scientific community will seldom or never embrace a new theory unless it solves all or almost all the quantitative, numerical puzzles that have been treated by its predecessor" ([1970], p. 20).

19. Collingwood (1956), p. 329; my italics. Elsewhere, Collingwood reiterates this claim: "Progress in science would consist in the supersession of one theory by another which served both to explain all that the first theory explained, and also to explain . . . 'phenomena' which the first ought to have explained but could not. . . . Philosophy progresses in so far as one stage of its development solves the problems which defeated it in the last, without losing its hold on the solutions already achieved" ([1956], p. 332).

20. Popper (1963). As he puts it elsewhere: "A new theory, however revolutionary, must always be able to explain fully the success of its predecessor. In all those cases in which its predecessor was successful, it must yield results at least as good . . ." ([1975], p. 83).

21. Cf. Lakatos (1970), p. 118.

22. Post (1971), p. 229. Cf. also Koertge (1973). Phenomenological theories of progress, every bit as much as positivistic and idealistic ones, are committed to the cumulativity postulate. For a detailed example, see Harris (1970), especially pp. 352–69.

23. Cf. especially Kuhn (1962), p. 169.

24. This was pointed out, among others, by Berzelius (1815).

25. Home's study (1972–73) makes it quite clear that Franklin realized this failure of his theory, but did not regard it as sufficient grounds for rejecting it. One might add that Franklin's theory also failed to give any solution *at all* for

the fact—widely observed and explained before his time—that there was generally a correlation between the density of a substance and its capacity to act as an electrical conductor.

26. We can illustrate what is involved by an example. Suppose that our scientific aim is to understand the embryology of birds. We have one theory, T_e, which offers a detailed account of the embryological development of eagles and egrets. We have another, T_s, which explains the embryological developments of all birds smaller than eagles, including egrets, but does not work for eagles. In such a circumstance, we would certainly view T_s as preferable to (i.e., a progressive improvement on) T_e, even if T_s was not able to solve the problem about embryonic development for eagles. Such a plausible judgment would be disallowed on almost all the standard (cumulative) theories of scientific progress. (For a fuller treatment of these issues, cf. L. Laudan [1976b].)

27. For the relevant discussions see Lakatos (1970), pp. 137, 175–77, and Kuhn (1962), pp. 11ff. and (1968).

28. Lakatos—at best—has shown how a programme could conceivably be progressive, while ignoring many anomalies; but that is a far cry from the stronger claim—required by his theory of mature science—that such anomaly-ignoring programmes are *ipso facto* more progressive than programmes which pay serious attention to their anomalies.

29. Given Lakatos' (Kuhnian induced) aversion to anomalies, he would probably have regarded this very feature of the dichotomy as a *bonus*. For those of us who do not share his views on the irrelevance of anomalies and criticism, however, such untestability must count as a serious liability.

30. It is worth pondering what motivates the search for a distinction between immature and mature science. My guess is that the quest harkens back to the old inductivist-positivist conviciton that "proper" science only began with Galileo, Newton, and the other classic heroes of the seventeenth century. Although eschewing inductivism, both Kuhn and Lakatos propose a demarcation criterion between mature and immature science which resurrects the inductivists' search for a definite point in time at which science became genuinely "scientific." (For a lengthy illustration of a historian's efforts to write about the history of science by utilizing such a demarcation criterion, see Gillispie's whiggish [1960].)

Chapter 5

1. Agassi (1963).
2. Grünbaum (1963).
3. For a guide to much of this literature, see Suppe (1974).
4. With the exception of Lakatos who is committed to this thesis. (See below, p. 165.)
5. Giere (1973).
6. *Ibid.*, p. 292.

7. *Ibid.*, p. 293.

8. *Ibid.*

9. *Ibid.*, p. 290.

10. Most philosophers of science ultimately do fall back on such a class of "privileged intuitions" about specific episodes as the final arbiter. Popper, for instance, writes: "It is only from the consequences of my definition of empirical science, and from the methodological decisions which depend upon this definition, that *the scientist will be able to see how far it conforms to his intuitive idea of the goal of his endeavors*" (my italics; Popper [1959], p. 55).

11. For a detailed exploration of those issues, see McMullin's valuable discussion (1970).

12. Cf. Lakatos' claims that: (1) "*All methodologies . . . can be criticised by criticising the rational historical recontructions to which they lead*" ([1971], p. 109); (2) "*A rationality theory . . . is to be rejected if it is inconsistent with an accepted 'basic value judgment' of the scientific élite*" ([1971], p. 110); (3) ". . . better rational reconstructions . . . can always reconstruct more of actual great science as rational" ([1971], p. 117); and, more explicitly, (4) "*Thus progress in the theory of rationality is marked . . . by the reconstruction of a growing bulk of value-impregnated history as rational*" ([1971], p. 118).

13. Although Lakatos tries to avoid this dilemma (saying that no theory of rationality "can or should explain *all* history of science as rational" [1971], p. 118), it follows inevitably from his method of ranking theories of rationality that the best such theory is that which "rationalizes" the largest part of the history of science.

14. The bulk of this section is concerned with the role of norms in the history of scientific *ideas*. The other main branch of the subject, the *social* history of science, likewise utilizes norms of rationality, but in different ways than the history of ideas. These issues are discussed below, pp. 184ff, 201ff.

15. Agassi (1963).

16. For a candid voicing of such anxieties, see Cohen (1974).

17. The "ahistoricity" of these philosophers is pointed out by McMillin (1970), Machamer (1973), McEvoy (1975), and Beckman (1971).

18. Although implicit in much of his work, this doctrine is most explicitly formulated in Lakatos (1971). The method of rational reconstruction began initially as philosophical technique for shedding light on the nature of rational deliberation and decision making. In its original conception, it involved postulating contrived and artificial cases of choice, which were deliberately simplified in order to get a handle on the case; these over simplified cases were then to be rendered more applicable to the actual situation by the gradual addition of complicating factors.

19. *Ibid.*, p. 91.

20. *Ibid.*, p. 106.

21. Similarly, Törnebohm claims, in his "rational reconstruction" of seventeenth century astronomy, that "historical accidents [*sic*] which affected the growth of this knowledge are not of interest . . . I will therefore take the

liberty of making a reconstruction of the historical development. The cast consists of two people whom I have invented . . ." ([1970], p. 79).

22. Lakatos, (1971), p. 107.

23. *Ibid.*

24. *Ibid.*, p. 106.

25. A similar example of the dubious historical relevance of the techniques of rational reconstruction can be found in Watson's book-length study (1966) of the downfall of Cartesianism. Watson's procedure is to define "a model of late seventeenth century Cartesian metaphysical system," whose weaknesses he proceeds to explore. Watson attributes the downfall of Cartesianism to the failure of this "model" system to come to terms with serious weaknesses it exhibited. What is curious is that Watson freely acknowledges that "none of the Cartesians . . . professed a system of exactly the sort" defined by his model ([1966], p. 29). Given that no actual Cartesian accepted the Watsonian reconstruction, Watson's lengthy analysis cannot explain why the genuine Cartesian philosophy was abandoned. Watson's discussion of the logical flaws in his imaginatively *ersatz* version of Cartesianism, for all its suggestiveness, never becomes authentic history.

26. Lakatos (1971), p. 107.

27. *Ibid.*, p. 108.

28. *Ibid.*, p. 107.

29. In fact, of course, there is not even strong similarity here, for the reconstructionist is not appraising the rationality of historical episodes, but feigned ones.

30. As already observed, it is probably the predilection of many "historically oriented" philosophers (from Hegel to Lakatos) for the cavalier method of rational reconstruction which makes most historians so suspicious of philosophical attempts to deal with the history of thought.

Chapter 6

1. See especially the discussion of conceptual problems in chapter two.

2. Kuhn (1968), p. 81.

3. *Ibid.*

4. Kuhn's beliefs about disciplinary autonomy are widely shared among historians, of both the "old" inductivist and the "new" socially oriented school. For references to some of the relevant literature, see below, pp. 213-17.

5. Hodge's study of the evolution of Lamarck's ideas (1970-71) exhibits vividly how important it is to attend to the problems a scientist is trying to resolve. Hodge points out that a widespread misconstrual of Lamarck's problem situation has led many historians to misinterpret the whole thrust of his theoretical research. (For a similar analysis of Chambers' work, cf. Hodge [1972].)

6. Cf. especially Gilson (1951) and Popkin (1960).

7. Compare Karl Jaspers: "The great philosophers . . . are best approached as contemporaries. . . . We shall understand them best by questioning them, side by side, without regard for history and their place in it" ([1962], p. xi).

8. This is not to say, of course, that there is nothing in common between the three. But historical understanding very often depends upon our ability to recognize that, in the course of time, problems undergo subtle, and sometimes profound, changes of both formulation and substance. As Quentin Skinner aptly remarks: "It is this essential belief that each of the classic writers may be expected to consider and explicate some determinable set of "fundamental concepts" of "perennial interest" which seems to be the basic source of the confusions engendered by this approach to studying the history of either literary or philosophical ideas" ([1969], p. 5).

9. Nelson writes: "The history of philosophy itself is the succession of increasingly successful solutions of these [unchanging] problems" ([1962], p. 22).

10. See especially Collingwood (1939).

11. *Ibid.*, p. 70.

12. The very possibility of answering such questions is denied by one of the more banal movements now popular in intellectual history, specifically, that form of structuralism associated with the work of Michel Foucault, especially (1970). For our purposes, the two chief flaws in Foucaultian historiography are: (a) *its completely stochastic character.* The "archaeology of ideas" (Foucault's version of intellectual history) offers no means, indeed denies the possibility, of ever giving a coherent acount of how world views ("epistemes") give way to one another, or of their mutual interconnections. Because Foucault insists that the emergence of new conceptual systems are the result of "ruptures of human consciousness," there can be no explanation—neither intellectual nor socio-economic—of the processes whereby new epistemes displace older ones. The next flaw is (b) *its vague invocation of the Zeitgeist.* Although allegedly eschewing traditional categories of historical analysis, Foucault's search for the common structures and metaphors which (on his view) permeate the thought of any epoch harkens back to the old, oft discredited belief that ideas "in the air" and "the collective consciousness" are the appropriate causal modalities for the historian. To understand a classic text, for Foucault, is neither to relate it to the biography of its author nor to examine the arguments within it; rather the historian studies such texts in order to find out what they tell us about the (linguistic) consciousness of an era. With its twin emphases on the mystery and the opacity of human thought, with its stress on "history as poetry," Foucaultian structuralism must rank as one of the most obscurantist historiographical fashions of the twentieth century. It says something about the state of mind of many intellectual historians that they are prepared to pay obeisance to a work like Foucault's which they generally concede to be unintelligible. Like Bergson and Teilhard before him, Foucault has benefited from that curious Anglo-American view that if a Frenchman talks nonsense it must rest on a profundity which is too deep for a speaker of English to comprehend.

13. I have tried to give some preliminary answers to these questions in L. Laudan (1973a), and (1977).

14. Cf. Holton (1973), especially chapters one and three. Holton claims to have identified most of the core concepts ("themata") ever to have occurred in the history of science and "suspect[s] the total will turn out to be less than 100" ([1975], p. 331).

15. See above, chapter three.

16. Skinner (1969).

17. These profiles correspond roughly to exegetical or descriptive history.

18. And they have been massively discredited *by experience.*

19. Cf. Lakatos (1963).

20. Popper is entirely typical in arguing that "in science (and only in science) can we say that we have made genuine progress: that we know more than we did before" (Popper [1970], p. 57).

21. It was one of Lakatos' genuine insights that this classical sacred cow of the philosophers had to be abandoned before one could develop an *adequate* theory of rationality. For examples of this approach, see Lakatos (1968) and L. Laudan (1973a).

Chapter 7

1. Although the bulk of this chapter will focus specifically on the sociology of knowledge, most of its conclusions also apply, *mutatis mutandis,* to the psycho-history of ideas.

2. For instance, unless a scientist believes in subatomic particles, he is hardly likely to join a laboratory doing research on the structure of the nucleus!

3. See, for instance, Scheler who asserts that "the sociological character of all knowledge, all forms of thought, intuition and cognition is unquestionable" (Quoted in Merton [1949], p. 231).

4. A propos these two extremes, it is more than a little ironic that Mannheim, who chastizes the "older" intellectual historians for making the *a priori* assumption "that changes in ideas were to be understood on the level of ideas" ([1936], p. 268), is himself committed—in what can only be described as an equally *a priori* fashion—to the view that virtually *all* changes in ideas are "bound up with social existence" *(Ibid.,* p. 278).

5. Mannheim grappled with this problem (unsuccessfully) through most of his career. On the one side, he wanted to insist that sociology had shown the social origins of virtually all systems of belief, including sociology itself: "Once we have familiarized ourselves with the conception that the ideologies of our opponents are, after all, just the function of their position in the world, we cannot refrain from concluding that our own ideas, too, are functions of a social position" ([1952], p. 145). On the other hand, as Mannheim gradually realized that such a view would vitiate the claims of sociology to possess

objective validity (and perhaps under pressure from the arguments of Alfred Weber), he began to argue that the thinkers—*such as himself*—were often immune from social influences and he developed the notion of "the relatively socially unattached intelligentsia" *(Ibid.,* pp. 252ff.). But if the intelligentsia can transcend social determination, and if the history of ideas is mainly concerned with the intelligentsia, what scope—even on Mannheim's account—is left to cognitive sociology?

6. Grünwald (1934), p. 229.

7. For an articulation of this distinction, see especially Mannheim (1936), chapter five.

8. It may be, of course, that the acceptance of beliefs *y* and *z* is a function of social factors, in which case we might say that the acceptance of *x* (rationally dictated by *y* and *z*) is *indirectly* the result of the social situation. But this does not controvert the claim that the most direct and most fundamental explanation for the acceptance of *x* by some thinker is that it follows rationally from *y* and *z*.

9. Merton (1949), pp. 516, 558. For Mannheim's formulation of this assumption, cf. (1936), p. 267.

10. It is as true today as it was when Mannheim pointed it out in 1931, that "the most important task of the sociology of knowledge . . . is to demonstrate its [explanatory] capacity in actual research in the historico-social realm" *(Ibid.,* p. 306).

11. For a discussion of some of these conceptual problems see below pp. 217ff.

12. A point similar to this was made by Imre Lakatos when he wrote, "*internal history* [of science] *is primary, external history* [of science] *only secondary, since the important problems of external history are defined by internal history*" ([1971], p. 105). What handicaps Lakatos' analysis is a failure to recognize the difference between cognitive and non-cognitive attempts to deal with the history of science. Although we are entitled to say that the "important problems" of *cognitive* sociology are, as it were, defined by the rational history of science, it is manifestly untrue to believe that the "important problems" of non-cognitive sociology are, to any significant degree, defined by the so-called internal (or rational) history of science.

13. See, for instance, Kuhn's remark quoted above p. 234 n. 40.

14. Richter (1973), p. 81; my italics.

15. *Ibid.,* p. 6.

16. Barber (1962).

17. Barber, for instance, speaks of Kelvin's "blindness" in opposing Maxwell's theory of light because the latter was not sufficiently mechanistic *(Ibid.,* p. 540). One may, with the advantage of hindsight, quibble with Kelvin's search for mechanical models; but in the historical circumstance there was nothing blind or irrational about Kelvin's initial reaction to Maxwell's work.

18. Mannheim effectively concedes this point in (1952), pp. 181f.

19. What I am calling contextual beliefs are more commonly called "existentially" or "situationally determined beliefs." I have avoided the latter terminology because it needlessly conjures up images of nineteenth-century German academic philosophy which are irrelevant to the case at hand.

20. Mannheim (1936), p. 272. Also, see pp. 265, 266, 271f.

21. Cf. Mannheim (1936), pp. 264–299 *passim.*

22. My italics; quoted by Merton (1949), p. 232.

23. For some examples, see below, pp. 220, 221.

24. If my inclusion of Durkheim seems peculiar, one need only recall his argument that anytime the acceptance or rejection of concepts is determined by their compatibility with prevailing beliefs, then we must be dealing with a "sociological process."

25. Brown (1970), p. 29.

26. Forman: "it was only as and when this romantic reaction against exact science had achieved sufficient popularity inside and outside the university to seriously undermine the social standing of the physicists and mathematicians that they were impelled to come to terms with it" ([1971], p. 110).

27. Compare, for instance, Kuhn's views about "the unparalleled insulation of mature scientific communities from the demands of the laity and of everyday life" ([1962], p. 163). See also my discussion of Kuhn's views on disciplinary autonomy above, pp. 173–75.

28. This belief that all intellectual conflicts and debates are in essence, a sublimated form of social conflict, permeates the work of many historians of science. As the social historian Steven Shapin puts it, the "good" historian must "try to assimilate conflict in ideas to conflict among competing groups in society" ([1975], p. 221). It is hard to regard this belief (or related ones such as "scientific disciplines are reactionary," "scientists only worry about philosophy when their prestige is threatened," "influences of the cultural environment on science must be caused by social factors," *ad nauseam)* as anything other than purely *a priori* prejudices, since none of the historians who subscribe to them ever offers even the pretense of a justification for them. (For a detailed critique of some of Shapin's views, see Cantor [1975b].)

29. Forman (1971), p. 6. Confronted by bald assertions of this kind, it is difficult to resist the *ad hominem* hypothesis that social historians are massively engaged in projecting their own disciplinary insecurities onto the history of science, convinced that scientists are as sensitive to questions of prestige as these historians evidently are.

This criticism is more than purely rhetorical. As Mannheim concedes, the entire discipline of the sociology of knowledge emerged as a generalization from the features of sociology itself. Early twentieth-century sociologists, examining the history of their *own* discipline, came to the conclusion that it was full of doctrines which owed more to the social background of their defenders than to their intrinsic rational merits. The general thesis of the sociology of knowledge (to wit, that ideas in most disciplines are socially determined) was founded on the hope that all other forms of knowledge might prove to be as subjective as sociology clearly was.

One sees this phenomenon in microcosm as well as in macrocosm by examining some of the more candid statements of working social historians of science. Steven Shapin, for instance, seeks to justify the reduction of scientific theory-choice to straightforward cases of social conflict by arguing that we usually seek in "everyday" life to explain "people's behavior and motives" ([1975], pp. 220–224) by reducing them to social causes, rather than paying attention to the reasons people give for their actions and beliefs. Can Shapin *really* believe that in "everyday" life, we *never* conceive that people believe things because they have good, nonsocial reasons for doing so? Can he be serious when he maintains that the social motivations of belief are "relatively familiar and known" when contrasted with the intellectual motivations for belief? In a different vein, Thackray (1970) urges that history of science must become more sociological and less intellectual in order to gain esteem in the eyes of general historians, sociologists, and campus radicals!

Virtually every conceivable reason for doing sociology of science has been rehearsed in the recent literature, *except* the argument that sociology might be able to offer some convincing explanations of important historical situations.

30. For all their opposition to "whiggish history," and to looking at the past through the spectacles of the present, Kuhn, Forman, and Brown are all guilty of projecting into the past a conception of disciplinary autonomy and insularity which derives from generalizations about present-day science. No conscientious survey of seventeenth, eighteenth, or nineteenth century science could have produced the Kuhn-Brown-Forman view that, as Forman puts it, "when scientists and their enterprise are enjoying high prestige . . . they are also relatively free to ignore the specific doctrines . . . which constitute the corresponding intellectual milieu" ([1971], p. 6).

31. It is revealing that when Forman's sociological model fails to explain the beliefs of scientists (as he admits it does in certain cases), he insists that we must look for some "psychological" explanation for why a scientist resisted the social forces upon him, rather than looking for some rational account of the scientist's belief. (Cf. especially Forman [1971], pp. 114-115.)

32. Consider, for instance, Elkana's recent claim that "the law of conservation could not be born either within the institutional framework of France or that of England" ([1974], p. 155). What are the general rules or laws of sociology which would warrant such a sweeping assertion? Where are the detailed case studies of the relation between institutional frameworks and scientific discoveries which might make us reasonably confident that we understood enough about the circumstances in which theories emerge to be warranted in asserting claims as strong as Elkana's?

33. Cf. Hessen (1971).

34. Ben-David (1971), pp. 13–14.

35. Merton (1970), p. 75.

36. Mannheim (1952), p. 135.

37. Richter (1973), p. 6.

38. Mannheim (1936), p. 288.

39. *Ibid.*

40. Very similar conclusions apply to the psycho-history of scientific knowledge, which is probably even further from possessing a psycho-dynamical model which can correlate beliefs about the natural world with psychological (or psychiatric) dispositions. Questions about whether, say, manic-depressives tend to favor field theories are on about the same level as whether gentlemen prefer blondes!

Bibliography

Agassi, J. "Towards an Historiography of Science." *History and Theory* Beiheft 2 (1963).

———. "Scientific Problems and their Roots in Metaphysics." In *The Critical Approach to Science and Philosophy,* edited by M. Bunge, pp. 189–211, 1964.

Aiton, E. *The Vortex Theory of Planetary Motions.* London, 1972.

Barber, B. "Resistance by Scientists to Scientific Discovery." *Science* 134 (1961): 596ff. (My references are to this paper as reprinted in Barber, B., and Hirsch, W., eds. *Sociology of Science.* New York, pp. 539ff., 1962.)

Bartley, W. "Theories of Demarcation between Science and Metaphysics." In *Problems in the Philosophy of Science,* edited by Lakatos and Musgrave, pp. 46–64. Amsterdam, 1968.

Beckman, T. "On the Use of Historical Examples in Agassi's 'Sensationalism'." *Stud. Hist. Phil. Sci.* 1 (1971): 293ff.

Ben-David, J. *The Scientist's Role in Society.* Englewood Cliffs, New Jersey, 1971.

Berzelius, J. "Essay on the Cause of Chemical Proportions." *Ann. Phil.* 2 (1813):443ff.

———. "An Address to those Chemists Who Wish to Examine the Laws of Chemical Proportions." *Ann. Phil.* 5 (1815): 122ff.

Boring, E. "The Dual Role of the *Zeitgeist* in Scientific Creativity." In *The Validation of Scientific Theories,* edited by P. Frank, pp. 187ff. New York, 1961.

Brooke, J. "Organic Synthesis and the Unification of Chemistry—a Reappraisal." *Brit. J. Hist. Sci.* 5(1970–71): 363ff.

Brown, T. *The Mechanical Philosophy and the Animal Oeconomy.* Unpublished dissertation, Princeton University, 1968.

———. "The Electric Current in Early 19th-century French Physics." *Hist. Stud. in the Phy. Sci.* 1 (1969): 61ff.

———. "The College of Physicians and the Acceptance of Iatro-Mechanism in England, 1665–1695." *Bull. of the History of Medicine* 44 (1970): 12ff.

Brush, S. "A History of Random Process. I. Brownian Movement from Brown to Perrin." *Archive for History of Exact Sciences* 5 (1968–69): 1–36.

Buchdahl, G. "Sources of Skepticism in Atomic Theory." *Brit. J. Phil. Sci.* 10 (1959): 120–34.

———. *Metaphysics and Philosophy of Science.* London, 1969.

———. "History of Science and Criteria of Choice." In *Historical and Philosophical Perspectives of Science,* edited by R. Stuewer, pp. 204ff. Minneapolis, 1970.

———. "Methodological Aspects of Kepler's Theory of Refraction." *Stud. Hist. Phil. Sci.* 3 (1972): 265ff.

Bunge, M. *Scientific Research.* 2 v. Berlin, 1967.

Butts, R. "Consilience of Inductions and the Problem of Conceptual Change in Science," In *Pittsburgh Series in Philosophy of Science,* edited by R. Colodny, forthcoming.

Cantor, G. "The Changing Role of Young's Ether." *Brit. J. Hist. Sci.* 5 (1970–71): 44ff.

———. "Henry Brougham and the Scottish Methodological Tradition." *Stud. Hist. Phil. Sci.* 2 (1971): 68ff.

—. "The Edinburgh Phrenology Debate: 1803–1828." *Annals of Science* 32 (1975a): 195ff.

——. "A Critique of Shapin's Social Interpretation of the Edinburgh Phrenology Debate." *Annals of Science* 32 (1975b):245ff.

Carnap, R. *Logical Foundations of Probability.* 2nd ed. Chicago, 1962.

Cohen, I. B. "History and the Philosopher of Science." In *The Structure of Scientific Theories,* edited by F. Suppe, pp. 308ff. Urbana, 1974.

Collingwood, R. G. *Autobiography.* Oxford, 1939.

———. *The Idea of History.* New York, 1956.

Costabel, P. *Leibniz and Dynamics; the Texts of 1692.* Ithaca, New York, 1973.

Culotta, C. "German Biophysics, Objective Knowledge, and Romanticism." *Historical Studies in the Physical Sciences* 4 (1974): 3ff.

Duhem, P. *The Aim and Structure of Physical Theory.* Princeton, 1954.

Durkheim, E. *Elementary Forms of the Religious Life.* Glencoe, Illinois, 1947.

Elkana, Y. *The Discovery of the Conservation of Energy.* London, 1974.

Ellegard, A. "The Darwinian Theory and 19th-Century Philosophies of Science," *J. Hist. Ideas* 18 (1957): 360ff.

Eriksson, B. *Problems of an Empirical Sociology of Knowledge.* Uppsala, 1975.

Faraday, M. "An Answer to Dr. Hare's Letter on Certain Theoretical Opinions." *Phil. Mag.* 17 (1840): 54–65.

Farley, J. "The Spontaneous Generation Controversy, I & II." *J. Hist. Bio.* 5 (1972): 95ff., 285ff.

Feyerabend, P. "Problems of Empiricism." In *Beyond the Edge of Certainty,* edited by R. Colodny, pp. 145–260. Englewood Cliffs, New Jersey, 1965.

———. "Problems of Empiricism, II." In *The Nature and Function of Scientific Theory,* edited by R. Colodny. Pittsburgh, 1970a.

——. "Against Method." In *Minnesota Studies in the Philosophy of Science,* vol. 4. Minneapolis, 1970b.

——. "Consolations for the Specialist," in *Criticism and the Growth of Knowledge,* edited by Lakatos and Musgrave, pp. 197ff. Cambridge, 1970c.

——. *Against Method.* London, 1975.

Fischer, D. *Historians' Fallacies: Toward a Logic of Historical Thought.* New York, 1970.

Forman, P. "Weimar Culture, Causality, and Quantum Theory, 1918-1927: Adaptation by German Physicists and Mathematicians to a Hostile Intellectual Environment." *Historical Studies in the Physical Sciences* 3 (1971): 1ff.

Foucault, M. *The Order of Things.* New York, 1970.

Fox, R. "The Rise and Fall of Laplacian Physics." *Historical Studies in the Physical Sciences* 4 (1974): 89ff.

Frank, P. "The Variety of Reasons for the Acceptance of Scientific Theories." In *The Validation of Scientific Theories,* edited by P. Frank, pp. 13ff. New York, 1961.

Ghiselin, M. *The Triumph of the Darwinian Method.* Berkeley, 1969.

Giere, R. "History and Philosophy of Science: Intimate Relationship or Marriage of Convenience?" *Brit. J. Phil. Sci.* 24 (1973): 282-97.

Gillispie, C. *The Edge of Objectivity.* Princeton, 1960.

Gilson, E. *Études sur le rôle de la pensée médièvale.* Paris, 1951.

Goldberg, S. "Poincaré's Silence and Einstein's Relativity." *Brit. J. Hist. Sci.* 5 (1970-71): 73ff.

Grünbaum, A. "The Duhemian Argument," *Phil. of Sci.* 11 (1960): 75-87.

——. "The Special Theory of Relativity as a Case Study of the Importance of Philosophy of Science for the History of Science." In *Philosophy of Science,* vol. I, edited by B. Baumrin. New York, 1963.

——. "Can We Ascertain the Falsity of a Scientific Hypothesis," *Studium Generale* 22 (1969): 1061-93.

——. *Philosophical Problems of Space and Time.* 2nd ed. Dordrecht, 1973.

——. "Can a Theory Answer More Questions than One of Its Rivals?" *Brit. J. Phil. Sci.* 27 (1976a): 1ff.

——. *"Ad Hoc* Auxiliary Hypotheses and Falsificationism." *Brit. J. Phil. Sci.* 27 (1976b).

Grünwald, E. *Das Problem einer Soziologie des Wissens.* Wien, 1934.

Hare, R. "A Letter to Prof. Faraday on Certain Theoretical Opinions." *Phil. Mag.* 17 (1840): 44-54.

Harris, E. *Hypothesis and Perception.* London, 1970.

Heimann, P. "Maxwell and the Modes of Consistent Representation." *Archive for History of Exact Sciences* 6 (1969-70): 171ff.

Hessen, B. *The Social and Economic Roots of Newton's "Principia."* New York, 1971.

Hodge, M.J.S.P.H.D. "Lamarck's Science of Living Bodies." *Brit. J. Hist. Sci.* 5 (1970-71): 323ff.

——. "The Universal Gestation of Nature: Chambers' *Vestiges* and *Explanations.*" *J. Hist. Bio.* 5 (1972): 127ff.

——. "Methodological Issues in the Darwinian Controversy." Forthcoming.

Holton, G. *Thematic Origins of Scientific Thought.* Cambridge, Mass., 1973.

——. "On the Role of Themata in Scientific Thought." *Science* 188 (1975): 328ff.

Home, R. "Francis Hauksbee's Theory of Electricity." *Archive for History of Exact Sciences* 4 (1967–68): 203ff.

——. "Franklin's Electrical Atmospheres." *Brit. J. Hist. Sci.* 6 (1972–73): 343ff.

Hooykaas, R. *The Principle of Uniformity in Geology, Biology and Theology.* Leiden, 1963.

Hull, D. *Darwin and his Critics.* Cambridge, Mass., 1973.

——. "Central Subjects and Historical Narratives." *History and Theory* 14 (1975): 253ff.

Iltis, C. "The Leibnizian-Newtonian Debates: Natural Philosophy and Social Psychology." *Brit. J. Hist. Sci.* 6 (1972–73): 343ff.

Jaspers, K. *The Great Philosophers.* New York, 1962.

King, M. "Reason, Tradition, and the Progressiveness of Science." *History and Theory* 10 (1971): 3ff.

Knight, D. *Atoms and Elements.* London, 1970.

Koertge, N. "Theory Change in Science." In *Conceptual Change,* edited by Pearce and Maynard, pp. 167ff. Dordrecht, 1973.

Kopnin, P., *et. al.,* eds. *Logik der wissenschaflichen Forschung.* Berlin, 1969.

Korch, H. *Die wissenschaftliche Hypothese.* Berlin, 1972.

Kordig, C. *The Justification of Scientific Change.* Dordrecht, 1971.

Koyré, A. "Review of Crombie's *Robert Grosseteste.*" *Diogéne* no. 16. October 1956.

Kuhn, T. *The Structure of Scientific Revolutions.* Chicago, 1962.

——. "History of Science." In *International Encyclopedia of the Social Sciences,* pp. 74–83. New York, 1968.

——. "Logic of Discovery or Psychology of Research?" In *Criticism and the Growth of Knowledge,* edited by Lakatos and Musgrave, pp. 1ff. Cambridge, 1970.

Lakatos, I. "Proofs and Refutations." *B.J.P.S.* 14 (1963): 1–25, 120–39, 221–43, 296–342.

——. "Criticism and the Methodology of Scientific Research Programmes." *Proc. Aristotelian Soc.* 69 (1968a): 149ff.

——. "Changes in the Problem of Inductive Logic." In *The Problem of Inductive Logic,* edited by I. Lakatos, pp. 315–417. New York, 1968b.

——. "Falsification and the Methodology of Scientific Research Programmes." In *Criticism and the Growth of Knowledge,* edited by Lakatos and Musgrave, pp. 91ff. Cambridge, 1970.

——. "History of Science and its Rational Reconstructions." In *Boston*

Studies in the Philosophy of Science, vol. 8, edited by R. Buck and R. Cohen, pp. 91ff, 1971.

Lakatos, I., and Zahar, E. "Why did Copernicus' Research Program Supercede Ptolemy's?" In *The Copernican Achievement,* edited by R. Westman, pp. 354ff. Berkeley, 1975.

Laudan, L. "Grünbaum on the 'Duhemian Argument'." *Philosophy of Science* 32 (1965): 295ff. (Reprinted in S. Harding, ed. *Can Theories Be Refuted?* Dordrecht, 1976.)

———. "Thomas Reid and the Newtonian Turn of British Methodological Thought." In *The Methodological Heritage of Newton,* edited by Butts and Davis, pp. 103ff. Toronto, 1970.

———. "C. S. Peirce and the Trivialization of the Self-Corrective Thesis." In *Foundations of Scientific Method in the 19th Century,* edited by R. Giere and R. Westfall, pp. 275ff. Bloomington, 1973a.

———. "G. L. Le Sage: a Case Study in the Interaction of Physics and Philosophy." In *Logic, Methodology and Philosophy of Science-IV,* edited by P. Suppes *et. al.,* pp. 429ff. Amsterdam, 1973b.

———. "The Methodological Foundations of Mach's Opposition to Atomism." In *Space and Time, Matter and Motion,* edited by P. Machamer and R. Turnbull, pp. 390ff. Columbus, 1976.

———. "Two Dogmas of Methodology." *Philosophy of Science* 43 (1976b).

———. "The Sources of Modern Methodology." In *Logic, Methodology and Philosophy of Science-V,* edited by R. Butts and J. Hintikka, Dordrecht, 1977.

Laudan, R. "Ideas and Institutions: the Case of the Geological Society of London." *Isis,* forthcoming.

Leplin, J. "The Concept of an *Ad Hoc* Hypothesis." *Stud. Hist. Phil. Sci.* 5 (1975): 309–45.

Lukes, S. "Some Problems about Rationality." *Archives Européenes de Sociologie* 8 (1967): 247ff.

McEvoy, J. "A 'Revolutionary' Philosophy of Science." *Philosophy of Science* 42 (1975): 49ff.

McEvoy, J., and McGuire, J. "God and Nature: Priestley's Way of Rational Dissent." *Hist. Stud. Phys. Sci.* 5 (1975).

McGuire, J. "Atoms and the 'Analogy of Nature'." *Stud. Hist. Phil. Sci.* 1 (1970): 3ff.

McGuire, J. E., and Heimann, P. "Newtonian Forces and Lockean Powers." *Hist. Stud. in Phys. Sci* 3 (1971): 233ff.

Machamer, P. "Feyerabend and Galileo." *Stud. Hist. Phil. Sci.* 4 (1973): 1ff.

McKie, D., and Partington, J. "Historical Studies on the Phlogiston Theory, I-IV." *Annals of Science* 2 (1937): 361ff; 3 (1938): 1ff and 337ff; 4 (1939): 113ff.

McMullin, E. "The History and Philosophy of Science: a Taxonomy." In *Historical and Philosophical Perspectives of Science,* edited by R. Stuewer, p. 12ff. Minneapolis, 1970.

Mannheim, K. *Ideology and Utopia.* London, 1936.

———. *Essays on the Sociology of Knowledge.* London, 1952.

Martin, E. *Historie des monstres dupuis l'antiquité jusqu'á nos jours.* Paris, 1880.

Masterman, M. "The Nature of a Paradigm." In *Criticism and the Growth of Knowledge,* edited by Lakatos and Musgrave, pp. 59ff. Cambridge, 1970.

Maxwell, A. "A Critique of Popper's Views on Scientific Method." *Phil. Sci.* 39 (1972): 31–52.

Merton, R. *Social Theory and Social Structure.* Chicago, 1949.

———. *Science, Technology and Society in 17th-century England.* New York, 1970.

Mittelstrass, J. "Methodological Elements of Keplerian Astronomy." *Stud. Hist. Phil. Sci.* 3 (1972): 203ff.

———. *Die Möglichkeit von Wissenchaft.* Frankfurt am Main, 1974.

Mitroff, I. *The Subjective Side of Science.* Amsterdam, 1974.

Mutschalow, I. "Das Problem als Kategorie der Logik der wissenchaftlichen Erkenntnis." *Voprosy Filosofii* 11 (1964): 27–36.

Nelson, L. "What is the History of Philosophy?" *Ratio,* 1962.

Neurath, O. "Pseudorationalismus der Falsifikation." *Erkenntnis* 5 (1935): 353–65.

Nye, M. J. *Molecular Reality.* London, 1972.

———. "Gustave LeBon's Black Light: a Study in Physics and Philosophy in France at the Turn of the Century." *Hist. Stud. in the Phys. Sci.* 4 (1974): 163ff.

Olson, R. *Scottish Philosophy and British Physics, 1750–1880.* Princeton, 1975.

Oresme, N. *A Treatise on the Uniformity and Difformity of Intensities.* Edited by M. Clagett. Madison, Wisconsin, 1968.

Pepper, S. "On the Cognitive Value of World Hypotheses." *Journal of Philosophy* 33 (1936): 575–77.

Popkin, R. *The History of Scepticism from Erasmus to Descartes.* Assen, 1960.

Popper, K. *The Logic of Scientific Discovery.* London, 1959.

———. *Conjectures and Refutations.* London, 1963.

———. *Objective Knowledge.* Oxford, 1972.

———. "The Rationality of Scientific Revolutions." In *Problems of Scientific Revolution,* edited by R. Harré, pp. 72–101. Oxford, 1975.

Post, H. "Correspondence, Invariance and Heuristics." *Stud. Hist. Phil. Sci.* 2 (1971): 213ff.

Quine, W. *From a Logical Point of View.* Cambridge, Mass., 1953.

Rescher, N. *Methodological Pragmatism,* forthcoming.

Richter, M. *Science as a Cultural Process.* New York, 1973.

Roger, J. *Les sciences de la vie dans la pensée française du XVIIIe siècle.* Paris. 1963.

Rudwick, M. "Uniformity and Progression." In *Perspectives in the History of Science and Technology,* edited by D. Roller, pp. 209ff. Norman, Oklahoma, 1971.

Sabra, A. *Theories of Light from Descartes to Newton.* London, 1967.

Salmon, W. "Bayes's Theorem and the History of Science." In *Historical and Philosophical Perspectives of Science,* edited by R. Stuewer, pp. 68ff. Minneapolis, 1970.

Schaffner, K. "Outlines of a Logic of Comparative Theory Evaluation." In *Historical and Philosophical Perspectives of Science,* edited by R. Stuewer, pp. 311ff. Minneapolis, 1970.

——. *Nineteenth-century Aether Theories.* Oxford, 1972.

——. "Einstein vs. Lorentz." *Brit. J. Phil. Sci.* 25 (1974): 45–78.

Schagrin, M. "Resistance to Ohm's Law." *Amer. J. of Phys.* 31 (1963): 536–47.

Scheffler, I. *Science and Subjectivity.* Indianapolis, 1967.

Schofield, R. *Mechanism and Materialism.* Princeton, 1970.

Shapere, D. "The Structure of Scientific Revolutions." *Phil. Rev.* 73 (1964): 383–94.

——. "Meaning and Scientific Change." In *Mind and Cosmos,* edited by R. Colodny, pp. 41ff. Pittsburgh, 1966.

Shapin, S. "Phrenological Knowledge and the Social Structure of Early 19th-century Edinburgh." *Annals of Science* 32 (1975): 219ff.

Shapiro, A. "Kinematic Optics: A Study of the Wave Theory of Light in the 17th-century." *Archive for History of Exact Sciences* 11 (1973): 134ff.

Sharikow, W. "Das wissenschaftliche Problem." In *Logik der wissenschaftlichen Forschung,* edited by P. Koptin *et. al.* Berlin, 1972.

Simon, H. "Scientific Discovery and the Psychology of Problem Solving." In *Mind and Cosmos,* edited by R. Colodny, pp. 22ff. Pittsburgh, 1966.

Skinner, Q. "Meaning and Understanding in the History of Ideas." *History and Theory* 8 (1969): 3ff.

Sloan, P. "John Locke, John Ray and the Problem of the Natural System." *J. Hist. Biol.* 5 (1972): 1ff.

Stallo, J. *Concepts and Theories of Modern Physics.* Cambridge, Mass., 1960.

Stegmüller, W. "Theoriendynamik . . . ," *Theorie der Wissenschaftgeschichte,* edited by W. Diederich, Frankfurt am Main, pp. 167ff.

Suppe, F., ed. *The Structure of Scientific Theories.* Urbana, 1974.

Thackray, A. "Has the Present Past a Future?" In *Historical and Philosophical Perspectives of Science,* edited by R. Stuewer. Minneapolis, 1970.

Törnebohm, H. "The Growth of a Theoretical Model." In *Physics, Logic and History.* London, 1970.

Toulmin, S. "Does the Distinction between Normal and Revolutionary Science Hold Water?" In *Criticism and the Growth of Knowledge,* edited by I. Lakatos and A. Musgrave, pp. 39ff. Cambridge, 1970.

Truesdell, C. *Essays in the History of Mechanics.* New York, 1968.

Vartanian, A. "Trembley's Polyp, La Mettrie, and 18th-Century French Materialism." In *Roots of Scientific Thought,* edited by P. Wiener and A. Noland, pp. 497ff. New York, 1957.

Viner, J. "Adam Smith and laissez faire." In *Adam Smith, 1776–1926.* Chicago, 1928.

Watkins, J. "Influential and Confirmable Metaphysics." *Mind,* N.S. 67 (1958): 344–65.

Watson, R. *The Downfall of Cartesianism: 1673–1712.* The Hague, 1966.

Whewell, W. *The Philosophy of Inductive Sciences, Founded upon their History.* 2v., London, 1840.

———. *On the Philosophy of Discovery.* London, 1860.

Winch, P. "Understanding a Primitive Society." *Amer. Phil. Quart.* 1 (1964): 307ff.

Wittich, D., *et al.*, eds. *Problemstruktur und Problemverhalten in der wissenschaftlichen Forschung.* Rostock, 1966.

Zahar, E. "Why did Einstein's Programme Supersede Lorentz's? I, II." *Brit. J. Phil. Sci.* 24 (1973): 95ff., 223ff.

Index of Names

Agassi, 156, 164, 168, 238, 239
Aiton, 236
Ampère, 84, 105
Aquinas, 97, 131
Aristotle, 1, 2, 25, 36, 51, 55, 58, 97, 112, 131, 134, 160, 214
Arschloch, 245

Bacon, 26, 59, 180, 185, 207, 224
Barber, 207-208, 243
Barrow, 144
Beckman, 239
Ben-David, 219, 245
Bergson, 241
Berkeley, 46, 135
Bernard, 58
Bernoulli, 25, 99, 104, 135
Berzelius, 31, 234, 237
Biot, 105, 228
Black, 83
Boerhaave, 83
Bohr, 71, 169, 182, 232
Boltzmann, 84
Borelli, 83
Boring, 102, 103
Boscovich, 104, 135
Boyle, 41, 85, 180, 218, 219
Brewster, 19
Brongniart, 19
Brooke, 230, 231
Brown, R., 19, 228
Brown, T., 214-217, 233, 234, 244, 245
Brush, 228
Buchdahl, 62, 230, 231
Buffon, 231
Butts, 230

Cantor, 230, 231, 233, 244
Carnap, 4, 47, 227
Carnot, 23, 90, 91, 92, 94
Chambers, 240
Charleton, 217
Clapeyron, 94
Clarke, 63
Clasius, 23, 95
Cohen, 239
Collingwood, 47, 121, 147-150, 177, 178, 183, 186, 237, 241
Comte, 231
Condorcet, 147
Conybeare, 228
Copernicus, 21, 46, 47, 55, 110, 112, 141, 218, 229, 230, 232
Costabel, 236
Cotes, 62, 231
Cramer, 21
Culotta, 231
Cuvier, 148, 149

Dalton, 113, 117, 149, 234
Darwin, 2, 34, 46, 63, 75, 78, 97, 101, 117, 136, 137, 138, 141, 207, 218, 229, 231
Democritus, 182
Descartes, 25, 29, 34, 36, 47, 52, 58, 75, 79, 80, 81, 85, 87, 88, 89, 91, 95, 97, 99, 101, 103, 104, 105, 117, 127, 139, 144, 166, 176, 180, 218, 233, 240
Dilthey, 183
Duhem, 27, 40-44, 119, 147, 182, 229, 230, 231
Dujardin, 19
Durkheim, 212, 213, 214, 244

Eddington, 23
Einstein, 20, 23, 34, 48, 49, 71, 89, 127, 139, 141, 227
Elkana, 245
Ellegard, 231
Engles, 175
Euclid, 201
Eudoxus, 51
Euler, 135

Faraday, 49, 50, 87, 99, 181, 230
Feuerbach, 105
Feyerabend, 3, 4, 47, 66, 74, 110, 113, 141, 143, 148, 156, 168, 182, 227, 231, 234
Fischer, 171
FitzGerald, 117
Forman, 214-217, 234, 244, 245
Foucalt, 241
Fourier, 105
Frank, 230
Franklin, 34, 90, 149, 229, 237
Frege, 100
Freud, 2, 46, 71, 78, 97

Galen, 214, 218
Galileo, 24, 25, 29, 33, 55, 58, 94, 112, 164, 218, 238
Ghiselin, 231
Giere, 158, 159, 238
Gillispie, 238
Gilson, 176, 240
Gorsseteste, 131
Grünbaum, 26, 77, 114-115, 156, 229, 232, 233, 234, 235, 238
Grünwald, 201, 243

Hales, 92, 93
Hanson, 66, 141, 143, 156, 182
Hare, 50, 230
Harris, 237
Hartley, 60, 85, 135
Hegel, 105, 199, 240
Heimann, 62, 230, 231, 236
Heisenberg, 216
Hertz, 85, 87
Hessen, 219, 245
Hintikka, 4, 227
Hippocrates, 58
Hobbes, 81, 144, 180, 181
Hodge, 231, 240
Holton, 182, 242
Home, 229, 237
Hooke, 81, 144
Hooykaas, 231
Hull, 231, 234
Hutchinson, 135
Hume, 177
Hutton, 62, 105, 145, 148, 233
Huygens, 25, 46, 61, 64, 81, 85, 88, 89, 95, 97, 135, 144, 149, 232, 233

Iltis, 232

Jaspers, 241

Kant, 62, 135, 179
Kelvin, *see* Thomson
Kepler, 29, 35
Knight, 231
Koertge, 232, 237
Kordig, 237
Koyré, 58, 59, 230
Kramers, 71
Kuhn, 1, 3, 4, 37, 47, 66, 72-78, 96, 99, 100, 109, 110, 113, 133-136, 138, 141-143, 145, 147, 148, 150, 151, 156, 171, 174, 175, 206, 207, 215-216, 227, 228-234, 236, 237, 238, 240, 243, 244, 245

Lakatos, 4, 26, 47, 66, 72, 76-77, 78, 91, 95, 97, 99, 100, 113-115, 118, 129, 147-151, 155, 156, 163, 168-170, 191, 227, 228, 229, 232-240, 242, 243
Lamarck, 141, 218, 240
Lambert, 60, 135
La Mettrie, 21
Laudan, L. 229, 230, 231, 233, 235, 236, 238, 242
Laudan, R., 231
Lavoisier, 24, 94, 234
Leibniz, 46, 61, 63, 64, 88, 89, 104, 127, 135, 180, 182, 232
Lenin, 218
Leplin, 235
Le Sage, 60, 135
Locke, 46, 59, 175, 183
Lorentz, 116, 117, 232
Lovejoy, 181, 182
Lukes, 236
Lyell, 80, 81, 135-136, 148, 149, 218, 231, 233
Lyonnet, 21
Lysenko, 63

McEvoy, 230, 239
McGuire, 62, 230, 231, 236
Mach, 50, 58, 100
Machamer, 239
McKie, 232
McMullin, 156, 239
Malebranche, 130

Mannheim, 196, 201, 202, 209–213, 219–221, 242, 243, 244, 245
Margenau, 237
Martin, 228
Marx, 2, 63, 71, 78, 79, 80, 99, 101, 105, 132, 165, 178, 199, 218, 219
Masterman, 231
Maupertuis, 104, 135
Maxwell, A., 235
Maxwell, J., 71, 85, 87, 94, 117, 230, 243
Mendel, 35, 63
Mersenne, 218
Merton, 202, 219, 220, 242, 243, 244, 245
Michelson, 34, 87
Mill, 26, 177, 179, 209, 210
Molland, 228
Morgan, 196
Morley, 34, 87
Musgrave, 232

Nelson, 177, 241
Neurath, 27, 229
Newton, 2, 23, 24, 25, 29, 34, 46, 47, 49, 52, 58, 59, 60, 61, 62, 64, 75, 79, 81, 84, 85, 88, 89, 93, 95, 97, 99, 100, 101, 103, 104, 105, 117, 127, 135, 137, 139, 141, 144, 146, 149, 160, 166, 178, 180, 181, 183, 188, 218, 219, 230, 231, 232, 233, 238
Nye, 228

Olson, 230
Oresme, 228

Parmides, 125
Partington, 232
Pasteur, 94
Peirce, 125, 147, 179, 223
Perrin, 20
Pitcairn, 83
Plank, 94, 182
Plato, 1, 51, 125, 182
Playfair, 233
Poisson, 105
Popkin, 176, 240
Popper, 4, 9, 26, 36, 47, 77, 114, 115, 124, 129, 136, 147, 148, 149, 227, 228, 229, 232, 235, 236, 239, 242
Post, 148, 237
Priestley, 62, 231, 234
Prout, 31, 169
Ptolemy, 24, 46, 47, 51, 52, 112, 117, 123, 141

Quine, 27, 141, 182, 229

Rayleigh, *see* Strutt
Régis, 81
Reichenbach, 4, 47, 125, 147
Richter, 207, 214, 220, 243, 245
Roger, 231
Rohault, 81
Rumford, *see* Thompson

Sabra, 230, 232
Salmon, 4, 45
Schaffner, 118, 233, 235
Scheffler, 129, 236
Scheler, 214, 242
Schofield, 104, 234
Shapere, 74, 231, 236
Shapin, 244, 245
Shapiro, 232, 233
Simon, 11
Skinner, B., 46, 183
Skinner, Q., 241, 242
Slater, 71
Smith, 105, 230
Sorokin, 214
Spengler, 215
Stahl, 97
Stallo, 50, 230
Stegmüller, 147
Strutt, 38
Suppe, 236, 238

Tarski, 77
Teilhard, 241
Thackray, 245
Thomas Aquinas, 97, 131
Thompson, 83, 149
Thomson, 57, 243
Törnebohm, 239
Toulmin, 70, 156
Trembley, 20
Truesdell, 236

van der Waals, 26
Vartanian, 20, 228
Viner, 230

Watkins, 232
Watson, 240
Weber, 218, 243
Wegener, 71
Werner, 145
Whewell, 50, 147, 156, 179, 230
Winch, 236
Wolff, 64

Young, 24, 88, 230

Zahar, 115, 118, 232, 235, 236